Industrial Hemp

as a Modern Commodity Crop

D.W. Williams, editor

American Society of Agronomy
Crop Science Society of America
Soil Science Society of America
5585 Guilford Rd., Madison, WI 53711-5801 USA

agronomy.org • crops.org • soils.org
dl.sciencesocieties.org
SocietyStore.org

Agronomy Monograph Series ISSN: 2156-3276 (online)
ISSN: 0065-4663 (print)

ISBN: 978-0-89118-631-1 (electronic)
ISBN: 978-0-89118-632-8 (print)

doi: 10.2134/industrialhemp

Cover design by Karen Brey
Cover photo credit: D.W. Williams

Table of Contents

A Preface to a Modern View of Industrial Hemp as a U.S. Commodity Crop   v

Chapter 1: The History of Hemp   1

Chapter 2: Hemp Grain   26

Chapter 3: Hemp Fibers   37

Chapter 4: Hemp Agronomy - Grain and Fiber Production   58

Chapter 5: Cannabinoids–Human Physiology and Agronomic   73
Principles for Production

Chapter 6: Hemp Genetics and Genomics   94

Chapter 7: Economic Issues and Perspectives for Industrial Hemp   109

Epilogue   122

A Preface to a Modern View of Industrial Hemp as a U.S. Commodity Crop

Doris Hamilton and D.W. Williams, Ph.D.

Growing and processing industrial hemp in the United States became federally legal again as a result of Section 7606 of the Agricultural Act of 2014, titled, "Legitimacy of Industrial Hemp Research". The Agricultural Act of 2014 is commonly called the 2014 Farm Bill and is codified as U.S.C. Section 5940, henceforth referred to as the Farm Bill. The Farm Bill allowed institutions of higher education and state departments of agriculture to grow industrial hemp for the purpose of research under an agricultural pilot program or other agricultural or academic research if such activities are allowed under the laws of the state.

The Farm Bill provided definitions to two critical terms, agricultural pilot program and industrial hemp. "The term 'agricultural pilot program' means a pilot program to study the growth, cultivation, or marketing of industrial hemp...." Additionally, the law defined "The term 'industrial hemp' means the plant Cannabis sativa L. and any part of such plant, whether growing or not, with a delta-9 tetrahydrocannabinol concentration of not more than 0.3 percent on a dry weight basis." Under this definition, industrial hemp was still a schedule I controlled substance under the definition of "marihuana" in the Controlled Substances Act (CSA: 21 U.S.C. 801 et seq.). Work began immediately in several U.S. states investigating allowable facets of hemp production and utilization under the new Farm Bill.

The U.S. Congress considers new Agricultural Acts (Farm Bills) on a 4-year cycle. Hence, the next consideration was in 2018. While the 2018 bill did not pass until very late in the year, it contained new language affecting hemp production and utilization in the U.S. Specifically, the bill contained language known as the Hemp Farming Act forwarded by Senate Majority Leader Mitch McConnell (R-KY) and supported by many including Representative James Comer (R-KY). The bill passed with the hemp language unaffected and was signed into law on 20 Dec 2018.

The 2018 Farm Bill changed the definition of industrial hemp by legalizing the derivatives from the plant, and also explicitly removed hemp from the definition of marihuana under the CSA, and doesn't include the word "industrial". Hence, hemp (as defined in the 2018 Farm Bill) is no longer a schedule 1 controlled substance under U.S. law. We note that on a federal level, marijuana (also written marihuana) remains a schedule 1 controlled substance under the CSA. This is true despite several states legalizing either the recreational and/or medical uses of marijuana.

The 2018 Farm Bill also directly provided for state regulation of industrial hemp production and processing activities under specific guidelines within the Bill. Even with these specific guidelines, the regulation of industrial hemp as an agricultural crop involves many unique challenges. Both industrial hemp and marijuana are Cannabis sativa L. The only legal difference between the two, according to federal law, is the measured concentration of THC. Industrial hemp and marijuana plants cannot be distinguished by physical differences. Neither the 2014 nor 2018 Farm Bills specified exactly how or when the delta-9 tetrahydrocannabinol (THC) is to be measured. Each state growing and processing industrial hemp is operating under different state laws, regulations, and policies. There are a great number of details to consider including: timing of the sample collection, portion of the plant to collect, analytical methods, and measurement of the delta-9 THC content pre- or post-decarboxylation. The 2018 Farm Bill did provide one additional stipulation regarding the testing by specifying that a state plan shall include "a procedure for testing, using post-decarboxylation or other similarly reliable methods, delta-9 tetrahydrocannabinol concentration levels of hemp produced". But, the general lack of clarity in federal requirements for sampling and testing has led to many differences among state programs, many of which still exist at this writing. There are even a few U.S. states that still consider hemp and its derivatives illegal. Conversely, there are also differences among states in the legal definition of hemp. Most states have adopted the 0.3% THC definition provided in federal law, but some states have different definitions up to 1% THC. This greatly complicates interstate commerce with hemp products.

As long as marijuana remains illegal on a federal level, hemp as a commodity crop must be strictly managed. Industrial hemp grain and fiber crops are grown using typical row-crop methods. Therefore, the resulting plant densities and cultivation techniques have a different appearance from typical marijuana production. However, the vast majority of industrial hemp production today is for the harvest of cannabinoids. Current cannabinoid production models are virtually identical to marijuana production. For this reason, it is important that state departments of agriculture who are administering hemp production and processing programs have policies in place to ensure the THC compliance of industrial hemp materials.

It is important to remember that state and federal law enforcement are tasked with enforcing all laws, including the laws governing the production of illicit cannabis. Many states have substantial production of illegal cannabis or marijuana, which makes the accurate identification of industrial hemp growing sites critical. Without accurate locations of licensed hemp growing sites, law enforcement would be required to waste valuable time and resources investigating industrial hemp sites. Also, without the proper identification of industrial hemp sites, it would be much more difficult for law enforcement to identify locations of criminal activity. Today, direct collaboration with law enforcement is imperative for the success of hemp as a commodity crop. If hemp is to become a broad-acre crop, distinct efforts must be made to ensure that industrial hemp production is not used as a cover for criminal activity.

The evolution of the hemp industry in the U.S. during 2014-2019 has been fascinating and in some ways unprecedented. In Kentucky, a state with one of the most active evolving industries, if we plot of the increase of participants and the associated acreage during the period 2014-2019, it is essentially an exponential function. There are several other U.S. states experiencing the same level of increased interest. One of the main drivers of this interest is cannabidiol, or CBD (see chapter 5). Today, gross income to farmers for CBD production is unprecedented; ranging from U.S $2000 to as much as U.S. $40,000 per acre. At the high end, this is clearly not a sustainable economic model. As are all commodity values in modern times, the ultimate value of CBD will be determined by the classical premise of supply and demand. When supply exceeds demand, the value will be adjusted appropriately in a negative direction. Today, we don't know when that will happen or to what degree the correction in price will be, but at this writing, many propose that the bubble will burst sooner than later.

A key determinant of the future of CBD as a commodity will be the regulation of the molecule by the U.S. Food and Drug Administration (FDA). The 2018 Farm Bill also included language that explicitly provides for full oversight of the cannabinoids, including CBD, to the FDA. The FDA released a public statement on the same day the 2018 Farm Bill was signed into law. The statement reiterated earlier opinions from the federal government regarding the lack of and distinct need for clinical research on the administration of cannabinoids to people and animals. As there are very few science-based, refereed reports on the effects of CBD and other hemp-derived compounds on humans and animals, many would agree that appropriate clinical evaluations are desperately needed before the cannabinoid molecules are freely administered to the public. What are the effects on developing nervous systems, either in otherwise healthy children, breast-feeding infants, or developing fetuses? What are other potential negative effects from regular ingestion of CBD, i.e., hepatic function and/or interactions with other drugs? What are the quantifiable positive effects, and at what doses are they achieved? There are many questions that must be addressed by solid scientific evaluations and clinical studies.

The FDA and U.S. Drug Enforcement Administration (DEA) approved one formulation of CBD as a schedule V controlled substance for the treatment of severe seizure disorders in 2018. At this writing, all other formulations of CBD are unauthorized by the FDA. The CBD industry anxiously awaits the public release of the FDA's decisions regarding the ultimate regulatory framework under which cannabinoids will be managed, as the 2018 Farm Bill provided full oversight to the FDA. This will literally determine the scope and scale of the U.S. cannabinoid industry. For example, if CBD and other cannabinoids are managed as schedule V controlled substances, among other things, this means they are available only by prescription. Cost effective production of a prescription medication must be predictable (repeatable) and under strict quality control, which is not practical in field-scale production. This would mean CBD production would likely be indoors and not a broad-acre crop; i.e., more of a horticultural than agricultural crop. If FDA regulations provide for broad inclusion of the molecules in beverages, foods, and other health-related products for humans and animals, then assuming demand meets current projections, production will be broad-acre, and almost certainly as a row crop and not under the female clone, spaced-plant production models that are common today.

From a purely pragmatic perspective, we offer that the best-case scenario for industrial hemp as a modern U.S. commodity crop may be much more grounded in basic agricultural economics than the industry is today. Fiber and grain production in 2019 are essentially equally profitable as are the other common commodities like corn and soybeans. Additionally, hemp fiber production could replace corn (as a full-season crop) in a normal crop rotation, and hemp grain could very simply replace late soybeans (planted after wheat), or in some environments full-season soybeans. We understand these economics very well and this is exactly what is happening in states with businesses and infrastructure to process the crops. Processing capacity is the current bottleneck in hemp fiber and grain production. Consumer demand for natural fiber and hemp grain products will define ultimate investments in processing businesses and infrastructure. As noted above, cannabinoid economics are essentially undefined today, and are remain undefinable until the federal regulatory framework becomes known. If cannabinoids do become common ingredients for many products aimed at humans and animals, U.S. producers will be in an excellent position to provide the molecule as a broad-acre commodity, assuming production is supported by profitability and economics. Time will tell.

This work endeavors to provide an extensive history of hemp, current information on the utilization and production of the three basic harvestable components; grain, fiber, and cannabinoids, the genetics and genomics of hemp, and lastly, current information on the economics of the U.S. hemp industry.

Hemp is simply new to the U.S. agricultural and general economies. There is so much that is unknown today that will define the ultimate consumer demands for hemp-based products. One thing we have learned in our work so far is that generally speaking, hemp is not spectacular in many regards and relative to many other plant species (e.g., other bast fiber and oilseed species). But, it certainly is unique in its ability to produce relatively high levels of cannabinoids. Will hemp become a real commodity crop in the U.S.? Again, time will tell.

This work is dedicated to my partner, Linda Williams. A true equal in all ways, she makes me far, far better than I am without her.

Chapter 1: The History of Hemp

John Fike, Ph.D.

The saga of industrial hemp would be an interesting history if only consider-ing its rise, fall, and possible rise again as a crop of global significance. The growing recognition that *Cannabis* (here, our designation as a genus) has played a consequential role in human development– and arguably has been an important crop for the advancement of humanity– makes its story that much more compel-ling. To both ends, this chapter sets out to explore the current understanding of industrial hemp's origins and its likely interactions with (and development by) humans as our species moved from living in roving hunter-gatherer clans to set-tled agrarian societies.

To clarify for the reader, this chapter will use *Cannabis* in its taxonomic sense (capitalized and italicized as a genus), while the term 'cannabis' will be used to designate the crop material undifferentiated by end use. 'Hemp' will linguis-tically delineate cannabis used as a food and fiber resource. Finally, the term cannabis also will be used in the brief discussion of the plant's history as an intoxicant– both for the sake of simplicity and to recognize the potentially undif-ferentiated use of the plant in our past.

Although this book largely is about hemp biology and agronomy, some discus-sion of the archeological literature is presented to explore humanity's long history with the plant. Interested readers will find more detailed discussion of such work in several books on the subject as well as reviews of the literature. And, while the focus of this chapter is on the history of cannabis as hemp, a food and fiber crop, abbreviated exploration of its use as an intoxicating and medicinal plant is war-ranted given the profound influence that these attributes have played in the genus' global colonization and domestication. Hemp has a fascinating history in terms of its past impact on human society and now has the potential to do so again in our future. As such this chapter aims to trace hemp's use by and relationships with societies (primarily in the West) through its rise, fall, and possible rise again.

Origins of *Cannabis*

The geographic origin of *Cannabis* has been the subject of long-running debate (e.g., see Clarke and Merlin, 2013, Small, 2015). Several locations for the species' nativity have been proposed, with various theories supporting Central, East, and South Asia (Clarke and Merlin, 2013, Liu et al., 2017, Mukherjee et al., 2008).

J. Fike, College of Agriculture and Life Sciences, Virginia Tech, Blacksburg, VA.
*Corresponding author (jfike@vt.edu)

doi:10.2134/industrialhemp.c1
© ASA, CSSA, and SSSA, 5585 Guilford Road, Madison, WI 53711, USA.
Industrial Hemp as a Modern Commodity Crop. D.W. Williams, editor.

Central and East Asia are considered more likely as the birthplace of *Cannabis* and would have allowed for the wide distribution of the species. A South Asian origin hypothesis is challenged by the difficulty the species would have faced to move north or south over the Himalaya and Hindu Kush mountain ranges (Clarke and Merlin, 2013). However, regions outside the point of origin likely were important in serving as glacial refugia for the species (Clarke and Merlin, 2013) during periods of unfavorable climatic conditions.

Humid temperate steppe biomes are thought to have provided the conditions for *Cannabis'* evolution (Gepts, 2004), and most experts consider that cannabis arose out of such conditions in Central Asia (Schultes, 1970, Small, 2015). A Central Asian origin also would have facilitated the wide early dispersal into Asia and Europe that followed as humans began interacting with the plant, although such arguments are not definitive (Clarke and Merlin, 2013). Moreover, recent assertions of an East Asian origin based largely on historical evidence of *Cannabis'* use (e.g., Liu et al., 2017) suggests that this argument (or perhaps claim for the species) is not settled. However, initial DNA analysis has provided support for the idea of a Central Asian genesis (Mukherjee et al., 2008), although cultivation of the crop likely began further east in what is present-day China and spread west from there (Li, 1973, 1974).

Centers of origin (now more typically called centers of diversity) are the ancestral regions that gave rise to the crop species humans have domesticated. Teasing out centers of diversity for a plant species typically involves determining where it grows in its wild form. Such centers are delineated in part by the fact that the crop ancestors in these regions have high degrees of genetic diversity. N. I. Vavilov, the eminent Russian scientist who developed these concepts, considered *Cannabis* to have had three centers of origin (Vavilov, 1951, cited by Clarke and Merlin, 2016). Large-seeded, broad-leaved fiber types (designated *C. sativa*) were thought to have originated in China; narrow-leaved narcotic types (designated *C. indica*) in India and Pakistan; and, a narrow-leaved type also designated *C. indica* (grown for seed) in Central Asia and the Tian Shan Mountains (Clarke and Merlin, 2013). This has been part of the basis for identifying the ancestral "home" for

Cannabis, but Clark and Merlin (2016) note that the assumption that high plant diversity reflects a center of origin may not be appropriate. The observed diversity may simply reflect "derivative instead of ancestral" variation, or variation due to human interaction compared to natural evolution.

In the case of *Cannabis*, it is more likely that these "centers" are less points of origin and more indicators of early human agriculture, with differences among the types reflecting the regional and human selection pressures for food, drugs, or fibers (Clarke and Merlin, 2013). Present debates over the geographical origins are in part a function of the long association that cannabis has had with humans. However, the degree of human interaction with and the corresponding flow of genetic material between cannabis the crop and any extant wild populations of the genus makes finding truly ancestral forms of the plant extremely unlikely (Meijer et al., 2003, Small, 2015).

As noted above, both species of *Cannabis* may contain narrow-leaved and broad-leaved forms, with each having both low- and high-intoxicant types. Among those "splitters" who separate *Cannabis* into different species, the narrow leaved form is more frequently designated *C. sativa* and the wider-leaved drug type *C. indica*. This terminology is confounded by the fact that scientists and those who use "sativa" and "indica" as colloquial appellations often have different meanings for these designations (Sawler et al., 2015). Additional confusion sometimes arises from the fact that a putative third *Cannabis* type, *C. ruderalis*, was identified in Central Asia by D.E. Janischewsky, a colleague of Vavilov's (Janischewsky, 1924, cited by Hillig and Mahlberg, 2004). The *C. ruderalis* form is shorter and shrubbier and some have considered it a wild ancestor of cultivated *Cannabis*, but chemotaxic evidence instead has indicated that it is a form of *C. sativa* (Hillig and Mahlberg, 2004).

Brief Consideration of Cannabis Speciation and Taxonomy

The challenges of piecing together the historic origins of *Cannabis* and detangling its different taxonomic relationships are intimately linked. Until very recently, genetic analyses have been limited due the inaccessibility of the plant to scientists. Thus, the debates about *Cannabis* and the broader treatment of species within the Cannabaceae are likely to continue for some time to come as the scientific community, long prohibited from easily researching the plant, is just beginning to explore the species with the full complement of genetic analytical tools. Still, an assortment of scientific approaches and historical records across a variety of disciplines have been applied to the question of *Cannabis*' speciation– albeit often with different conclusions.

Using evidence from archaeological, paleobotanical, and palynological records and genetic studies, Clarke and Merlin (2013) rendered up a detailed hypothesis of how *Cannabis* could have emerged as genus with multiple species. The authors concluded that *C. indica* and *C.sativa* should be considered separately and that based on the evidence "*C. indica* cultivars are the most geographically widespread and most widely utilized biotypes today, growing on all continents and used for recreational and medicinal drugs as well as fiber and seed production, while *C. sativa* cultivars are presently grown only for fiber and seed on limited acreage in Europe and North America" (Clarke and Merlin, 2013).

Others take a narrower view of *Cannabis* taxonomy (*e.g.*, Rahn et al., 2016, Small, 2017). Small (2015) noted that "no other species has generated so much misunderstanding, argument and contradictory literature", and has argued for a single taxon, *C. sativa*, with subspecies and varieties based on the biochemical nature of the plant material. This stance was taken because fiber and narcotic types are reproductively compatible (with a great deal of hybridization among them) and with the consideration that observed differences between the types largely reflect divergent selection pressures during domestication. Noting the challenge that taxonomists can face when considering whether to formally recognize a group of organisms (and if so, at what taxonomic rank), Small observed that "those who have espoused... recognition of more than one species...have done so without addressing the theory and practices of classification" (Small, 2015).

It is not the purpose of this chapter to engage in the arguments about speciation– that will be left to those whose expertise is in the precise definition of the genetics of the plant. For our purposes (and for simplicity), this chapter will follow the convention that all *Cannabis* is *C. sativa*. Interested readers are guided to Small (2015; 2017) for detailed considerations of the plant's taxonomy and comparison with other taxonomies, along with a suggested framework for describing subspecies based on plant morphology and chemistry. A contrasting vision of speciation based on a geographic history of the plant is put forward by Clarke and Merlin (2013).

Cannabis Prior to Human History

The family Cannabaceae is thought to have arisen during the Cretaceous period and contains 10 genera (Sytsma et al., 2002, Yang et al., 2013). Cannabaceae includes both *Cannabis* and *Humulus* (common name, hops), two relatively closely related genera. While *Cannabis* is an upright plant and *H. lupulus* is a vine, both utilize similar habitats and produce resinous material from secretory glands. Both genera also bear "seeds" (achenes, technically; an achene is a dry fruit containing a single seed) and have pollen quite similar in form (Small, 2017).

Palynology, the study of spores and pollen grains, often has been used to gain insight into ecosystems of the past. Along with allowing scientists to identify a species' presence or absence, palynology may provide a window into historical climatological and ecological conditions. Palynological studies have provided important insights into the history of *Cannabis* because aside from pollen grains, the fossil record does not provide evidence (and thus accurate dating) to the time of its evolution (Small, 2017).

The reproductive characteristics of *Cannabis* lend themselves well to palynology. Large female plants can bear hundreds of flowers and a single male flower produces hundreds of thousands of pollen grains (Fægri et al., 1989; cited by Small, 2015). The earliest reported evidence of *Cannabis* in the palynological record comes from core samples taken in the East European Plain. These samples indicate the presence of *Cannabis* in Eurasia as early as 150,000 yr ago (Molodkov and Bolikhovskaya, 2006). Readers should note, however, that older studies attempting to ascertain the timing and presence of *Cannabis* in the landscape based on the palynological record should viewed with caution because *Cannabis* and *Humulus* pollen are

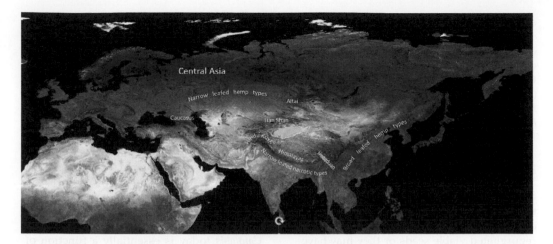

Fig. 1. Proposed regions of natural origin for Cannabis include Central Asia (between the Caucasus and Altai Mountains), South Asia (largely in the Himalayan Mountains) and East Asia in the Hengduan-Yungui region (Clarke and Merlin, 2013). Although some consider the Caucuses Region hemp's likely region of origin, each of these regions likely played important roles in Cannabis evolution and domestication. A central challenge for understanding hemp's origin arises from the intermingling of the different types, their advance and retreat in the face of a changing climate, and their ultimate use, transport and development by humans. Central Asia and Europe fostered largely narrow-leaved hemp types that were carried to North and South America for fiber production. Broad-leafed hemp types were common to East Asia and cultivated for food and fiber. A pocket of broad-leafed narcotic strains is localized to the Hindu Kush. Narrow leaved narcotic strains were more common around the Himalayas and spread from South Asia to Africa and on to the Western Hemisphere and narrow-leafed hemp strains reached the Western Hemisphere from Europe. More detailed discussion of these issues and of cannabis' migratory patterns is presented in Clarke and Merlin (2013).

quite similar in appearance (Fleming and Clarke, 1998). A recent (2013) summary of Cannabis pollen discoveries reported from around the world suggests a large gap exists in the Eurasian palynological record (or at least in the literature) in the period between 150,000 and 10,000 BP (Clarke and Merlin, 2013). More frequently, the bulk of early evidence of hemp pollen in the palynological record dates from around 10,000 BP (Clarke and Merlin, 2013). The palynological evidence for presence beginning from around that time likely reflects the advent of human interventions with the plant, but prior to the advent of human interaction Cannabis surely must have had some way of expanding out from its points of origin.

Conditions for Growth and Means of Distribution

While the palynological record delivers evidence for a site's ecological history, the distribution of today's wild or feral plant populations provide strong indicators of the climatic and edaphic conditions from which Cannabis would have evolved and been adapted. In this regard, Cannabis grows best at rather moderate temperatures– between about 59 °F and 81 °F (15 °C and 27 °C)– which could be expected for Cannabis in pre-history as well (although modern cultivars can tolerate quite low temperatures (Ehrensing, 1998). Edaphically, hemp is best suited to well-drained soils with high fertility and has little tolerance for waterlogged or poorly drained sites (Clarke and Merlin, 2013). The need for nutrients and moisture on well-drained sites also points to origins within river valleys along stream banks, lakeshores, and alluvial fans. It is telling that in North America, feral hemp often carries the moniker "ditchweed", linguistic evidence of such site preference.

Prior to humanity's interactions with Cannabis, its primary vectors of distribution likely would have been moving streams, birds, and mammals. Although strong winds might transport seed, this would be an ineffective mechanism because the small round shape of the seeds is not well suited to windborne dispersal. Birds may have been particularly important to the spread of *Cannabis* as many species readily consume the seeds, and some avian species have appellations taken from the name of the plant. *E.g.*, the species designation of the common (or sometime "hemp") linnet (*Linaria cannabina* L.) in Europe is derived from *Cannabis*, and field sparrows (likely *Passer montanus* L.) in China often are called 'hemp bird' (　; Clarke and Merlin, 2013).

Although birds may break seeds in the process of consumption, some species swallow seeds whole, and these can be stored in the bird's crop prior to processing and further digestion (e.g., Darwin, 1869, Zheng et al., 2011). In such cases, birds may have regurgitated viable seeds or they may have been killed by predators, with the seed dispersed before they were processed and digested. Darwin (1869) observed that birds (with seeds in their crops) could be blown several hundred kilometers off their course and noted that tired birds were subject to predation by raptors. Moreover, he reported that seeds of *Cannabis* (and other plant species) had germinated following residencies of up to 21 h in the stomachs of birds of prey.

Large ungulates likely also were important as vectors of *Cannabis*. Equids and cattle are known consumers of grains, including hemp. Equid mobility is such that they could disperse seeds over many kilometers in a day's time (Hampson et al., 2010). Although the fossil record is mute on this point, it is plausible that now-extinct ungulates and megafauna from hemp's home range would also have eaten and distributed hemp seeds over some distance. While a majority of seeds likely would lose viability during passage through the digestive tract, the smaller seeds of wild hemp are variable in hardness and germination (Small, 2015). Passage of viable seeds has been described in a number of situations with horses and ruminants (Aper et al., 2014; Nishida et al., 1998, Quinn et al., 2008, Rahimi et al., 2016), and colonization post excretion could have occurred even with a low percentage of excreted seed being viable.

Evidence of *Cannabis* in Human Pre-history

Human interaction with *Cannabis* likely began well before the evidence is available in the archaeological record. Some have even speculated that early humans carried seeds of the plant with them as they retreated from the last ice age 50,000 to 70,000 years ago, although proof for this is lacking (Clarke and Merlin, 2013). The point in time at which *Cannabis* entered human consciousness as a preferred species will always be ambiguous, but it was likely early in our history. Certainly, our nomadic ancestors would have recognized the plant's versatility as a food and fiber resource and probably as a medicinal or psychotropic herb used in shamanistic healing rituals. Interaction with humans would give the adaptable invader new opportunities for expansion, and the global distribution of *Cannabis* today is essentially a function of humanity's historic use of the species.

Early on, the sites where conditions were best suited for *Cannabis* likely would have been along trails and by the dung piles left by animals, particularly those near water sources and drainage ways. As humans used these trails and made new encampments, they would have created areas of disturbance, all the while depositing their own wastes and leaving trash heaps. Thus, the process of habitat expansion for humans would have created disturbed sites with elevated nutrient levels that would have been ideally suited for the plant given its preference for sunshine and well-drained sandy-loam soils of high fertility (Johnson, 1999, Schultes, 1970, Small, 2015, USDA, 2000). Because humans and plants utilized the same sites along streams and lakes, the two species likely would have been in frequent contact, facilitating *Cannabis'* collection, use, and spread during its early days as a "camp follower" (Schultes, 1970, Small, 2015).

At what stage in the course of human development *Cannabis* transitioned from being a weedy opportunistic vagrant to an actively collected and nurtured crop is another open question, but such entry to the path of domestication is common. In this regard, *Cannabis* is no different from "probably the majority of the world's major domesticated crops (which) are related to, or are known to have originated from (weedy, opportunistic) plants" (Small, 2015, see also Harlan, 1965). Early utilization could have been due in part because it was readily accessible (Clarke and Merlin, 2013), and Small (2015) conjectured that "it was almost certainly associated with humans in very early times".

Cannabis and the Pollen Record in the Age of Humans

Palynological studies detail the presence of *Cannabis* in Europe from after about 10,000 BP, coincident with human activities. Pollen records indicate *Cannabis'* presence in East Asia from about to 7000 BP (Japan) to 4500 BP (China) but the evidence is limited to a handful of studies (Clarke and Merlin, 2013). Earlier dates in China would be expected given our knowledge of the plant's long use in the region. Similarly, of the more than two dozen palynological studies that (Clarke and Merlin, 2013) cataloged, none gave evidence for *Cannabis* in South Asia. This is surprising since *Cannabis* had migrated to South Asia by around 3000 BP, carried south and west from Central Asia by nomads and traders (Fleming and Clarke, 1998). A strong pollen signal would be expected given the plant's well-recognized history of use within the region, but absence of evidence is not evidence of absence, and the lack of record may simply reflect limited research on this subject.

The palynological record indicates that *Cannabis* spread throughout Europe from west to east over northern and southern routes (Fleming and Clarke, 1998). *Cannabis* pollen found in Italy was dated to the late Pleistocene (Mercuri et al., 2002) while pollen finds along the northern route to Europe are signposts that *Cannabis* had reached the Baltic region by at least around 7600 BP (Poska and Saarse, 2006). Larger pollen peaks appear from the time of the Roman Empire (Fleming and Clarke, 1998) and testify that hemp played a role in the dramatic landscape-scale changes humans made to European ecosystems during the Iron Age (*e.g.*, Cyprien et al., 2004). Pollen finds closer to the present and associated with lake sediments indicate when and where Europeans had begun practicing water retting (described in **A history of Hemp as a Fiber Crop**-*Hemp Fiber Production and Processing*) for fiber processing (Andresen and Karg, 2011, Brombacher and Hecker, 2015).

Cannabis Seed Finds

Of the plant constituents (pollen, seeds, fibers) found in the archeological record that could testify to early *Cannabis* use, seeds provide the strongest direct witness. Fibers of *Cannabis* purportedly have been found at many archeological sites but they present more challenges for definitively determining the species (as

mentioned below). Seeds, in contrast, provide clear indication of past use.

The earliest seed-based evidence that humans used *Cannabis* comes from sites found in Japan (Kudo et al., 2009, Okazaki et al., 2011). The seeds were found along with pot shards which also had cord markings, giving circumstantial evidence that *Cannabis* may have been used as a fiber resource about 10,000 BP by the Jōmon people. Although seeds and cordmarked pottery do not definitively speak to *Cannabis* use as a fiber material, in this case such findings do present additional evidence of even earlier plant–human interactions. We know this because *Cannabis* is considered to have migrated east out of Central Asia, making its way with humans via land bridge to Japan about 18,000 BP (Clarke and Merlin, 2013). More recent seed-based evidence from northeast Asia dated to about 2000 BP has been linked to hemp fabrication in the region (Jia, 2007).

In Europe, seed-based evidence has been found in two burial pits within the Danube Valley region of Romania. These sites, dated to about 4000 BP, each contained vessels with carbonized hemp seeds (Rosetti, 1959), suggesting that *Cannabis* may have been burned or smoked as a component of incense. Another interpretation is that these vessels were ritual food containers used in a feast for the dead (Sherratt, 1981, cited by Clarke and Merlin, 2016) which would provide evidence that *Cannabis* was being used for food. Interested readers are directed to Clarke and Merlin (2013) and Long et al. (2017) who provide several additional references for *Cannabis* seed finds in Asia and Europe that point to various food and ritual uses from the Holocene to modern times.

String, Cannabis, and Early Human Advance

The use of string, cloth, and cord predates the rise of civilization, dating from about 30,000 BP (Adovasio et al., 2007, Barber, 1992) and fiber technologies have been fundamental to the advancement of humanity. Converting fibers to string had profound effects on the advancement of early human societies because of the myriad ways in which this technology could be used. Some consider the development of textiles second only to the use of cereal grains in the founding of human culture (Li, 1973, 1974). Thus, it is worthwhile to briefly digress from the *Cannabis* story to consider what more broadly has been called the "string" or "fiber revolution" (Adovasio et al., 2007; Barber, 1992).

Although much of the study and description of prehistory is immersed in the language of stones (Adovasio et al., 2007), the advent of fiber technologies made possible a number of human advances– including, in many cases, the use of the stones themselves. Fiber technologies would have made it possible to create a number of tools: for example, to weaponize a stick with a sharp stone point (a spear) which could be jabbed or thrown at enemies and prey. Coupled with a bow "spring", fibers could be used to propel smaller weaponized sticks (arrows) over distances and with great speed and force. Fiber technologies also would support more nurturing uses. The ability to lash together materials for shelter, to strap on a child, or to create new modes of clothing or baskets for carriage; all would have provided opportunity to expand humans' range and habitat, by reducing labor needs and opening up new possibilities for discovery and invention.

More recent interpretations of the anthropological record suggests that fiber development and use had more profound effects on human advancement than did any technical progress in making weapons, scrapers, and other stone tools (Adovasio et al., 2007). This connection of fibers to human development also points to the critical role that women played in our advancement as a species. Women often have been given bit parts in narratives of our early history because the tools of their craft, degradable fibers, are the first evidence to be lost from the archeological record; however, they were the likely leaders and innovators in early fiber technology development (Adovasio et al., 2007);

Development of new fiber technologies accompanied and supported the transition from nomadic to settled lifestyles. The invention of mesh and line technologies provided humans the capacity to catch, trap, and hook fish and waterfowl, ultimately resulting in supplies of protein for the community that would have been more stable. This in turn would allow for more extended periods of settlement at a given site (Clarke and Merlin, 2013). As nomadic existence gave way to life in transhumant or permanent settlement along riparian transportation routes, such changes in lifestyle would free both temporal and cognitive resources. This would allow for new pulses of creativity and innovation– just as such changes continue do in the present

(Bertinelli and Black, 2004). Experimentation with and domestication of plants and animals and development of new agricultural technologies would have supported and reinforced this transition. Considered in this context, it is easy to see how cannabis might have been an essential crop for early human societies. A species capable of supplying seed for food, medicinal and psychotropic compounds for healing and religious ritual, and strong fibers for a variety of tools could be a key resource.

The role cannabis played as the fiber resource for textiles, nets, and cords at these early junctures in our history remains a mystery. Certainly many fiber sources would have been available and utilized by our ancestors. Although the preponderance of fibers found in the archaeological record have come from Asia (Clarke and Merlin, 2013), the evidence for *Cannabis* as the source material frequently has been "based on the geographical and historical contexts in which the fiber remains were found. In almost all cases, no actual laboratory identification of the plant fibers has been provided" (Clarke and Merlin, 2013). The lack of analysis may reflect the difficulty of correctly identifying cannabis fibers from among other potential bast fibers. Sources such as flax (*Linum usitatissimum* L.), ramie (*Boehmeria* spp.) or nettle (*Urtica* spp.) often were available to and utilized by our ancestors.

Some of the earliest confirmation of fiber use derives not from actual fiber remains but from the forms that they cast. Imprints of fibers on earthen floors, in clay pottery, and on bronze surfaces (which had been wrapped in cloth) all have been ascribed to cannabis (Fleming and Clarke, 1998). Indirect evidence in cordbased pottery in Central European sites have been dated to about 25,000 and 27,000 BP and were thought to be used as netting to capture birds and small game (Pringle, 1997). However, the source of the impressions has not been positively identified. Cord marks on 12,000-year-old pottery artifacts may have been made by, or intentionally *with*, cannabis fibers (Chang, 1968, cited by Clarke and Merlin, 2013) and slightly younger evidence of intentionally cordmarked pottery comes from the Jōmon site **Fig. 2**. in Japan (Kudo et al., 2009). The fact that seeds were found with pottery shards does not provide definitive proof that cannabis fiber was the source of the marks, but it likely increases one's confidence in such conjecture. As with other "signals" of

Fig. 2. Incipient Jōmon pottery (14th–8th millennium BCE). Tokyo National Museum, Japan. Figure in public domain.

Cannabis presence and use, Clarke and Merlin (2013) have provided a detailed discussion of the historical record based on fiber finds and a number of references for fiber-based findings reported in the literature.

As noted before, much of the record of early fiber use comes from East Asia. *Cannabis* provided the only herbaceous plant fibers for the region that stretched from what is now northern China into eastern Siberia (Li, 1974). Proto-Chinese peoples utilized these fibers for clothing and fabrics and the materials for spinning the strands into thread are a common constituent of settlement artifacts found in the region (Cheng, 1982, cited by Clarke and Merlin, 2013. In contrast, evidence of cannabis fibers in the South Asian archaeological record appear limited despite the plant's occurrence there (Clarke and Merlin, 2013). However, these strains more typically were the diminutive psychotropic forms unsuited for fiber production, and other plant resources such as kudzu (*Pueraria montana* [Lour.] Merr.), abaca (*Musa textilis* Née), jute (*Corchorus* spp.) and ramie (*Boehmeria nivea* [L.] Gaudich.) were readily available and likely preferred as sources of fiber (Fleming and Clarke, 1998).

History of Cannabis in Religious and Medical Traditions

Although the focus of this text is on the use of Cannabis for fiber and food, a description of the plant's history without mention of its use in religious and medicinal traditions would be remiss. The very brief treatment that follows is necessarily short because our purpose is merely to consider the evidence for human use of Cannabis and the manner in which these uses helped support the plant's global expansion. Curious readers also will find a range of books and reviews on the topic. Some may consider the use of Cannabis as illicit or at least anathema (aside from industrial purposes). Indeed, that has informed U.S. policy regarding the plant for nearly a century. However, the circumstances and conditions of human living prior to the advent of modern medicine warrant consideration and contextualization. Nature's pharmacopoeia, historically has been (and for some cultures remains) the essential source of medicines for healing and pain relief. The relationships between humans and psychoactive animals, plants, and fungi began early in our history and probably served as the inspiration for early religious experience and practice (Clarke and Merlin, 2013). Some hypothesize that humans specifically sought such materials when moving into new areas to rise above normal states of consciousness or to communicate with ancestors or other parts of the spirit world. Discovery of such effects from Cannabis likely happened early in humans' interactions with the plant, given that the seeds are surrounded by flower bracts rich in psychoactive resins **Fig. 3.** Although research on the adaptive purposes of cannabinoids is limited, some have considered that Δ9-tetrahydrocannabinol (THC; the intoxicating compound in these resins) has little apparent benefit for preventing plant disease or predation. As such, others argue its presence may be for another adaptive purpose– attracting humans or other animals– and thus principally a human artifact (Schultes and Hofmann, 1992).

Evidence for *Cannabis'* pre-historic medicinal use has been traced back to 4000 BC with carbon-14 dating techniques (Russo, 2004, cited by Warf, 2014) and some of the earliest pharmacopeias from China chronicle the plant's medicinal and psychological effects (Li, 1974). Both the Chinese and Vedic (Indian) texts that discuss cannabis trace from older oral traditions, suggesting much earlier awareness and use of the plant (Clarke and Merlin, 2013).

Early recreational, religious, and medicinal uses that developed in East and South Asia have been accepted or punished by various societies, depending on the religious and political hierarchy of the day. In China, cannabis was used in Daoist religious ceremonies until those customs fell from grace around the sixth century CE (Li, 1974). This decline corresponds with the rise of Confucianism, a philosophy that rejects the role of the spiritual in human life. In contrast with China, pressures against the use of cannabis were limited if even extant in India, and the plant continues to be used today as part of some Indian religious traditions (Clarke and Merlin, 2013).

Cannabis also may have a biblical connection. Benet (1975) posited that the term 'cannabis' can be found in Jewish scriptures but that the original wording was lost when the Bible was (mis)translated into Greek. Benet (1975) considered 'cannabis' to derive from the earlier Semitic terms 'kaneh' (hemp) and 'bosm' (aromatic) and suggested the biblical context surrounding 'kaneh-bosm' indicate that it was used both for fiber and for religious ceremony.

Cannabis also was known in the Middle East, perhaps 1000 years or more before the rise of Islam. Consumption of hashish (the dry cannabis resin) receives no mention in early Islamic texts and apparently did not warrant mention until the 10th century (Clarke and Merlin, 2013). The drug was compared with alcohol (which is prohibited in Islam), but attempts to quash its use typically failed, despite penalties for use that historically have been quite stiff. In time, Arab traders carried cannabis to North Africa and down the continent's eastern coast, where it was used for psychotropic and euphoric properties (Warf, 2014). These materials dispersed across the continent and eventually would make their way to the Caribbean with East Indian workers who migrated from British East Africa following the end of slavery (Clarke and Merlin, 2013).

In Western Europe, descriptions of cannabis from early Greek and Roman civilizations appear limited largely to fiber and medicine. The Greek scholar Herodotus indicated its use in ritual by Scyhtian nomads to the east (Dewey, 1914), and some suggest cannabis was used at Greek oracles to facilitate communication with the dead (Bremmer, 2002, cited by Clarke and Merlin, 2016). Other writings from Greece and Rome suggest it was used recreationally (Clarke and Merlin, 2013), although such interpretations of these writings (particularly from Herodotus) have been questioned (Duvall, 2014). Pope Innocent VIII's 1484 papal bull "Summis desiderantes", a proclamation against witchcraft, is viewed by some as a statement against the use of cannabis, but the papal bull (Pope Innocent VIII, 1484) makes no such mention of the plant.

Cannabis has a more recent history of use in Western medicine following its (re)introduction to the West from India in the late 19th century (Clarke and Merlin, 2013, Small, 2017). Use declined in the early decades of the 20th century, however, because obtaining consistent results from plant materials and preparations that had variable potencies proved difficult. Passage of the 1937 Marihuana Tax Act put further pressure on cannabis use and it was removed from U.S. Pharmocopoeia in 1942. Healers back to antiquity have treated a variety of ailments with cannabis and modern medical interest in the plant and its chemistries greatly increased following discovery of cannabinoid receptors (Devane et al., 1988). Efficacy and mechanisms of these treatments needs verification, and Western science and medicine are just beginning to tease out what our ancestors knew or sensed about the medicinal properties of the species.

Cannabis in our Language

Along with the various artifacts and historical records, one can hear the echo of human interactions as facilitated by *Cannabis* sounding through our different languages. Etymology, the study of word origins and their changes over time, can provide clear evidence of different cultural exchange and interaction among peoples. Plant names are important for etymological study because typically they belong to ancient word groups (Ignatov et al., 2010) and can be used to test similarities among languages, thus providing linguistic linkages of peoples back across space and time. The similarities of words among different peoples can help reveal the origins and travels of humans and plants alike. Original terms for fiber (*cana*) and psychoactive (*bhang*) lines have understandably evolved over time with movement of the crop from region to region and people to people. Those interested in the linguistics of

Fig. 3. A medicinal cannabis plant in flower has a glossy appearance due to resin-laden flower bracts (inset). Figures in the public domain.

psychotropic cannabis will find such treatment in Clarke and Merlin (2013). A brief primer on the linguistics of fiber cannabis follows here.

The Chinese term for hemp, *má*, is depicted as (Li, 1974). Although hemp as food grain would be displaced by other crops by the 10th century (see the next section on hemp grains and oils), its appellation, *'má'* remains embedded in the Chinese language as the radicle (or root) of several other words. Words such as grind (hemp + stone) and porridge (hemp + rice) indicate the practical agronomic relationships of humans and hemp. *Má* also is the root for 'narcotic', 'numb', 'tangle' and 'troublesome' among other words, pointing to its psychotropic and medicinal effects. Similar linguistic relationships are found in the Korean language, indicating the various properties of cannabis were well known to the peoples of northeast Asia.

As noted above, the Latin 'cannabis' may trace to *kaneh bosm* from early Semitic languages which still can be heard in the Turkish *nasha* and Arabic *kannab* today (Benet, 1975). The term śā a or *cana* (from Sanskrit) likely gave rise to *kenab* (Farsi/

Persian) and these words have links to *kannabis* (Greek), *konopli* (Russian) and *konopj* (Polish). *Cañamo* (Spanish), *cânhamo* (Portuguese) and *chanvre* (French) all derive from Old Latin (Benet, 1975, Dewey, 1914). In turn, connections among Northern European languages may be seen in *cainb* (Gaelic), *hanf*, (German), *hennup* (Dutch) *hampa*, (Danish), *hamp* (Swedish) and *hemp* (Dewey, 1914).

In modern English, the crop's broad use and representative nature for long fibers also resulted in something of an inverse phenomenon etymologically. That is, the word 'hemp' has been used (and likely misconstrued) as a generic name for several other long fiber sources including Indian hemp (or jute), sisal hemp (*Agave sisalana* Perrine), and sunn hemp (*Crotalaria* spp.) among others (Dewey, 1914). Of course, 'cannabis' also gave rise to "canvas", reflecting this important use of hemp fibers (Harper, 2019).

Hemp Grains and Oils in Historic and Modern Contexts

Historical Uses of Hemp as a Food Crop

Scholars consider Northeastern China the cradle of hemp cultivation and suggest the first harvests from wild hemp plants occurred many as 8500 years ago (Li, 1974). Active cultivation likely began from about 6000 BP (Chen et al., 2009), when hemp was among a handful of foundational grains that included millets (*Setaria italica* (L.) P. Beauv. and to a lesser degree *Panicum miliaceum* L.) and buckwheat (*Fagopyrum esculentum* Moench) (Lee et al., 2007, Li et al., 2010, Li et al., 2009, Yang et al., 2012). Classical Chinese literature such as Shi Ching (Book of Odes) and Li Chi (Book of Rites) from about 2100 or 2200 BP provides some of the earliest written evidence and instruction on growing the crop for fiber and grain (Li ,1974b). The subsequent domestication and introduction of upland rice (*Oryza sativa* L.), soybean (*Glycine max* (L.) Merr.) and probably wheat (*Triticum aestivum* L.) and barley (*Hordeum vulgare* L.) (Lee et al., 2007) led to hemp's long-term decline as a food crop in the Far East. Although scholars differ on the timing, consumption as a grain staple in China likely ended sometime in the first millennium CE (Keng, 1973, Li, 1974). Today, Chinese citizens consume the grain as snacks, in geriatric diets, or as medicinal foods (Clarke and Merlin, 2013). Archeological records, and in some cases, modern consumption by peasants, suggest hemp seed likely was a routine dietary constituent for the peoples of Korea and Japan as well as those in Central and Southwestern Asia (Clarke and Merlin, 2013).

Europeans also consumed hemp, although the historical evidence for this use appears more limited. Traditional European hempseed recipes point to its long use from centuries past (Leson, 2013) and Clarke and Merlin (2013) provide a number of anecdotes and resources that point to its use as a food grain. While peoples across European social strata probably consumed hempseed (Clarke and Merlin, 2013), the crop likely more frequently sustained those living in poverty or served as a food resource in times of famine (Small, 2017).

Hemp produced specifically as a grain crop has precedent, although it historically occurred only within Russia (Small et al., 2007), unless birdseed production (in Canada, Europe and the United States [see Dewey, 1914]) also qualifies as such. Dewey (1914) tested Russian landraces and described these experimental plants as short, with compact seedheads that could be harvested similar to other seed crops, but which had little value for fiber. To date, most of the crop improvement efforts intending to yield grains have been geared toward "dual purpose" varieties that also could yield fibers. Short-statured, short season (60 to 90 d) grain cultivars have been developed for more northern latitudes (Callaway, 2003, Small and Marcus, 2002) and this is likely to be an area of research effort if hemp rejoins modern cropping systems.

New Consideration for Hemp as a Feedstuff

Despite the past association of hempseed with poverty feeding, renewed interest in hemp as a grain crop for feeds and food products has arisen largely out of the evidence of hemp seed's quantifiable nutritional value. Hemp seeds may serve as a source of protein and are relatively energy-dense, typically containing between 27 and 38% oil (Bócsa et al., 2005, Callaway, 2004, Kriese et al., 2004, Oomah et al., 2002, Vonapartis et al., 2015). Use of hemp oil both as food and fuel has been described in Chinese texts from about a millennium ago (Whitfield, 1999). Just as with hemp use

as food staple, consumption of hemp-based food oils in China fell from favor over time. Other seed crops had preferred oil qualities, although hemp remained a superior feedstock for lamp oil (Clarke and Merlin, 2013).

Interestingly, some practices described in the early Chinese texts, pressing the seeds for food and fuel oils and feeding the remaining press "cake" to fatten livestock– show that for hemp, the past may indeed be prologue. Feeding such residuals following the primary extraction of seed oils is a common practice in modern livestock production systems. Early studies show that hempseed cake is suitable for use in this manner (Hessle et al., 2008, Karlsson, Finell et al., 2010).

Researchers also have tested the feeding of full fat hemp seed to cattle, but performance results have been mixed. For example, including hemp seed at up to 14% of dietary dry matter animal did not affect growth metrics. Meat fatty acid profiles had both increased trans and saturated fats (generally considered a negative response) but increased conjugated linoleic acid (Gibb et al., 2005), an important dietary anticarcinogen. Other reports of altered milk chemistry and meat fatty acids (Cozma et al., 2015, Mourot and Guillevic, 2015) suggest that feeding hemp oil and seeds to livestock may have positive consequences for the human end consumer due to improvements in nutritional profiles. Although some have had concerns over animal products potentially having THC levels above safe standards (EFSA, 2011), this is unlikely to be of much concern (EFSA, 2015).

Given the value of the oils (see next section), future use of hemp products in animal feeds may be limited to the byproduct cake produced during extrusion. Still, adding such cakes to animal diets has potentially positive human nutrition implications. In one study, calf weight gain did not differ between hemp- and standard soybean- and barley-based diets (Hessle et al., 2008), but both greater concentrations of mono- and polyunsaturated fatty acids and a better n-6/n-3 ratio were reported for the hempseed-fed steers (Turner et al., 2008). The higher level of polyunsaturates in muscle tissues (Woods and Forbes, 2007) may represent a storage and handling issue, however, due to their greater potential for oxidation. Hempseed cake also has been fed in dairy diets with variable effects on milk yield, but quality responses in terms of fatty acid profiles were not reported.

Some suggest hempseed oil content and fatty acid profiles stand to give hemp strong market value and may create the opportunity for the plant's renaissance as a grain crop (*e.g.*, Small, 2017). However, while hempseed oil may prove beneficial to humans as food and nutritional supplements, such uses alone would only create opportunity for hemp in specialty or niche markets. Entry into the much broader food and feed system may prove challenging if hempseed cannot compete on price and value when pitted against traditional grain commodities.

Hemp Seed and Essential Oils for Human Food and Consumer Products

Historically, hempseed oil was used in a number of industrial products such as paints and varnishes (Small, 2017). These applications may again be revived as part of the 'green' (or not) products industries, but given current production scales, hemp's potential value likely will be greater in products for human use or consumption, that is, for foods, soaps, supplements, and cosmetics (Small, 2017). Such uses may well dwarf potential markets for use for industrial purposes and animal feedstuffs (Leson, 2013).

The recent history of demand for 'natural foods' in U.S. markets suggest significant future opportunities exist for hemp and have propelled the resurgent interest and efforts for grain production (Leson, 2013). Growth in this market is not surprising given that hemp seeds are considered a functional food. As noted previously, hempseeds have positive nutritional profiles and the fatty acid profile may be particularly nutritious. Hemp has a 3:1 linoleic to α-linoleic ratio, considered optimal for human health (Oomah et al., 2002) and γ-linolenic acid, which is not found in other major food grains. The seeds also have relatively high levels of vitamin E, insoluble fiber, and an array of minerals (Oomah et al., 2002, Small, 2015).

For centuries, humans have used dietary hempseed to treat various disorders (Callaway, 2004; Woods and Forbes, 2007). While full of promise as a nutraceutical food, the current excitement should be tempered by the fact that research regarding its benefits is limited and somewhat variable. Further, much of the work has been conducted with animal models. For example, in a comparison with fish oil, flaxseed oil, and hempseed oil,

research found hempseed oil had no effects on plasma fatty acids in healthy adults over a 12-wk period (Kaul et al., 2008), and none of the treatments affected platelet aggregation or inflammatory markers. In contrast, research with rabbits (*Oryctolagus cuniculus*), indicated that animals fed elevated levels of cholesterol had normal platelet aggregation when supplemented with hempseed meal (Prociuk, Edel et al., 2008). Hempseed meal hydrolysates also have been found useful in maintaining blood pressure of hypertensive rats (*Rattus* spp.). In a small human study, subjects that consumed hemp oils had better serum high density lipoprotein (HDL)-to-total cholesterol ratios relative to those who consumed flaxseed oil (Schwab et al., 2006).

More work is needed to verify various claims about hemp's efficacy for addressing any number of ailments. Compounds derived from various parts of the plant are being used in treatments as diverse as hypertension and oxidative stress (Girgih et al., 2014a, b) to inflammatory bowel disease (Parian and Limketkai, 2016), to cancer (Pathak et al., 2016). In addition to grains, hemp inflorescences stand to be a good source of essential oils (to include terpenes and cannabinoids) for medicinal compounds (Fernández-Ruiz et al., 2013) as well as for flavorings and fragrance additives (Bertoli et al., 2010). Hemp terpenes have moderate antimicrobial and insecticidal activities (Górski et al., 2009, Novak et al., 2001, Thomas et al., 2000) and a raft of studies have begun to evaluate cannabinoid efficacy against a number of diseases (Coetzee et al., 2007, Fernández-Ruiz et al., 2013, Radu and Robu, 2014, Rieder et al., 2010, Vera et al., 2012, Woods and Fearon, 2009).

A History of Hemp as a Fiber Crop

Fibers and Empires

As described previously, humanity's use and production of hemp fibers goes deep into our history. Evidence of hemp fiber in Central Asia dates to at least 4000 BCE (Li, 1974). Hemp-based fiber technology is probably much older and would have been shared through trade and invasion. These technologies would spread westward from Asia, likely carried by Thracians and

Scythians across Eurasia to lands east of the Caspian Sea (Dewey, 1914). Although the Scythians probably brought hemp to Europe around 1500 BCE, the plant did not receive recorded mention until about 450 BCE by Herodotus, perhaps suggesting the plant largely was unknown to the Greeks and Romans prior to that time (Dewey, 1914). Subsequent Roman conquest and colonization continued hemp's spread into Europe and the Mediterranean during the early centuries of the Common Era. After the fall of the Roman Empire, hemp's move across Western Europe carried on as various peoples mastered the techniques required for its production and use. The plant reached Northwestern Europe by around 1000 CE, if not before, evidenced by canvas beginning to replace woolen sails in Norway (Clarke, 2002); adoption in the region was further facilitated by Norse trade (Duvall, 2014).

Fibers that provided strong, durable textiles and cordage were essential to national empires built on feats of military, industrial, and agricultural engineering and to personal empires built on commerce. As such, hemp became a critical commodity in Europe. Countries, city-states, and merchant networks sought to control its production and use as a means to accumulate power and wealth (Duvall, 2014) and this played out in the interaction of the social hierarchy. Those interested in the social aspects of cannabis production and use (both as fiber and psychotropic) and its relationship to social power structures in historic and present-day contexts will find absorbing discussions in Warf (2014) and Duvall (2014).

The demand for hemp fibers, important for rope and fabrics in medieval Europe, expanded significantly, as nations developed navies and advanced sailing technologies. For example, hemp fibers were a critical resource for the Venetian city-state (Duvall, 2014), which dominated much of the eastern Mediterranean during its existence during the 13th to 19th centuries. Venice imported the raw fibers it needed from the surrounding region, and a vigorous state-controlled industry developed to process them into textiles (Duvall, 2014). This helped guarantee needed supplies for its navy, the source of the city-state's power.

Few European countries produced sufficient supply to meet their needs, despite hemp's being widely distributed and cultivated in Europe. A flourishing fiber production industry did develop around the Balkans (largely in present-day Germany, Poland, and Baltic states) that supplied hemp to parts of Europe from the 13th to 17th centuries, and political and social changes allowed Russia to enter these markets in the 18th century (Duvall, 2014). Many countries would supply their navies with high quality, low cost fiber from the Baltic, but routinely struggled with their reliance on foreign resources.

The rise of England, Spain and Portugal as maritime powers particularly increased the need for hemp imports into Western Europe. Sailing ships required tons of fiber to yield the yards of sails and the miles of line and cordage which powered and secured them. By the 1700s, a man-o-war might carry over 65 km (40 mi) of rope in active service and storage **Fig. 4.** As with Venice, however, these seafaring nations did not produce sufficient fiber to meet their own demands and their efforts to produce hemp economically met with variable (and usually limited) successes. Given the problems of being reliant on hemp from the Baltics, these countries made development of colonial hemp sources a priority.

Hemp as a New World Fiber Crop

Fiber hemp first arrived in the New World in the 1500s, brought by the Spanish to Mexico in the 1530s and to Chile in 1545 and by the Portuguese to Brazil by the 1600s (Campos, 2012, Duvall, 2014, Husbands, 1909). Spanish and Portuguese efforts to grow hemp both at home and in the colonies were driven by their sailing fleets' needs for cordage and cloth. Although Spain did grow some hemp, each country ultimately became dependent on Russian supplies. In Chile, a small, local industry developed based on production of rough fiber and seed and still exists today, but efforts to grow hemp in Mexico largely proved fruitless (Duvall, 2014). The Spanish also subsidized hemp production in present-day California during the late 1700s and early 1800s, but farmers largely stopped growing the crop when the subsidy ceased, coincident with the Mexican War of Independence.

France and Great Britain were the primary contributors to industrial hemp's production in Colonial-era North America. As with Spain and Portugal, each country's

Fig. 4. Images of the USS Constitution. Rigging and sails were critical for the development of commercial and naval fleets. Images in the public domain.

needs for fiber largely were driven by the demands of naval and commercial sailing fleets, which required many tons of rope and cordage for each ship. Along with efforts in North America, British colonists also attempted to grow hemp in South Africa, Australia, and New Zealand, which would all would prove fruitless (Duvall, 2014).

France was better positioned than Britain to meet its own fiber needs from within Europe, but at times its policies either encouraged or discouraged French colonists from growing hemp. French encouragement of hemp production began in Québec and Nova Scotia and met little success as colonists there found the crop unprofitable (Duvall, 2014). Growers apparently had better success in Louisiana; however, in the early 1700s, France prohibited commercial hemp production in the territory to protect its industry in the homeland (Gray and Thompson, 1933). The French would again encourage hemp production in the mid-1700s; second-hand accounts suggest that by the century's end, New Orleans had active ropeworks producing quality cordage (Gray and Thompson, 1933). Whatever hemp industry had developed proved short-lived, however, and by the time of the Louisiana Purchase (1803) the region's primary fiber was cotton (Duvall, 2014).

North American Hemp in the Colonial Era

Early efforts to grow hemp in the English colonies of the Atlantic Seaboard appeared more promising. Within a decade after the Virginia Company founded the Jamestown settlement in 1607, John Rolfe reported that Virginia's hemp and flax crops were as good as those in England and Holland (Gray and Thompson, 1933). In 1619, hemp production became "compulsory for all (Virginia) colonists having sufficient seed" (Gray and Thompson, 1933). Often the available seeds from Europe were no good due to poor storage during the trans-Atlantic transit, (Duvall, 2014; Herndon, 1963). Because of this, shortages of seed was a routine complaint, likely because farmers felt compelled to get seeds only from reliable neighbors (Herndon, 1963). New England colonists would try the crop around 1645 (Small and Marcus, 2002), and governments in 10 of the 13 colonies ultimately made policies to encourage hemp production (Duvall, 2014).

Despite these inducements and the crop's potential productivity, agronomic and economic conditions largely rendered hemp a crop for domestic use and it was never exported in large quantities (Gray and Thompson, 1933). Hemp required large amounts of nutrients, and this field preparation was no small task. For example, Virginia farmers typically worked the seedbed three times (fall and spring plowing and pre-plant harrowing) Herndon (1963). As well, harvesting and processing were extremely laborious, backbreaking and sometimes dangerous routines, as is discussed later. The labor demands for successful hemp production also put it in competition with production of food and cash crops, and the limited pool of labor was a frequent complaint of early colonial landowners (Duvall, 2014, Gray and Thompson, 1933). High labor costs, coupled with high freight charges and inferior processing methods, meant that colonial hemp generally was of lower quality and more expensive than the hemp reaching western Europe from the Baltic region (Gray and Thompson, 1933). Although the colonists were reported to have skilled textile producers (Gray and Thompson, 1933), trade law prevented shipping finished products to Europe (Duvall, 2014). Thus, early plantation owners generally shied away from growing the crop as a commodity.

Along with these competitive disadvantages, hemp as commodity crop likely faced a bigger challenge. The historical evidence suggests that the value of tobacco was a particular impediment to large-scale hemp production. This changed whenever tobacco prices fell, but growers reverted to tobacco production as soon as the market rebounded (Gray and Thompson, 1933).

Although limited as a commodity, hemp had wide usage domestically (Herndon, 1963) Weaving technologies introduced in the early 1700s allowed colonists to provide cordage and textiles for themselves and not be tied to expensive and variable finished goods imported from European markets. Thus, prior to the American Revolution the crop was widely grown, if not produced as a commercial crop. The revolutionary period was a notable exception (Herndon, 1966) Hemp production increased due to the lack of available British goods and was preferred to tobacco as a means of purchasing supplies during the war (Gray and Thompson, 1933; Herndon, 1963; Herndon, 1966).

Hemp as a Fiber Crop Following the American Revolution

As the United States expanded, settlers carried hemp further into North America, and a commercial cordage industry developed and flourished in Kentucky after 1775. This would assist the decline in eastern hemp production, and the industry would spread west to Missouri and Illinois through the mid-1880s (Roulac, 1997, cited by Fortenberry and Bennet, 2001).

Ironically, the rise of another fiber crop, cotton (*Gossypum hirsutum* L.) would support hemp's commercial success (Duvall, 2014). Hemp producers in Kentucky supplied the cotton industry with the textiles and cordage needed to bag and bale the crop. Geopolitics also were important for the new industry, as Europe's Napoleonic wars in the early 1800s helped raise the value of Kentucky hemp to the point that it was a staple crop for the state by 1810. Although the industry suffered once European hemp imports rebounded and hemp was subject to large fluctuations in value, it remained the principle market crop in Kentucky for several decades and would grow in importance in surrounding states (Gray and Thompson, 1933). However, in the 1840s and 1850s, some Kentucky growers began changing to alternative crops due to competition, limited labor, and deteriorating soils (Gray and Thompson, 1933).

In 1855, Kentucky hemp growers also suffered stand losses due to poor weather and in turn tried a variety of seed sources to reestablish the crop (Duvall, 2014). During this period, highly productive Chinese hemp cultigens were introduced and became the preferred plant material for fiber cropping (Duvall, 2014). Maintaining a supply of domestic seed was problematic, however, as the taller Chinese material frequently crossed with feral hemp strains; seed imports from China thus remained a necessity (Duvall, 2014). Thus, the term "Kentucky hemp" does not represent a cultivar, but a "conceptual centrality of (Kentucky's place) in U.S. hemp history" (Duvall, 2014).

In 1859, Kentucky and Missouri still produced over three fourths of U.S. hemp (Gray and Thompson, 1933), but war and its consequences would again play a role in the crop's future during the 1860s. Secession of the Southern states during the U.S. Civil War ushered in the beginnings of hemp's decline. Producers in the Midwestern (Union) states could no longer sell their fibers to their principle market, Southern cotton growers (Duvall, 2014). That market, in turn, already had an interest in using other means of binding cotton and the industry shifted to metal ties following the war (Duvall, 2014). Government subsidies for hemp production also ceased during the Civil War, as, thankfully, did the right to hold humans in bondage. Following war's end, newly freed men and women had much less compulsion to return to the toils and suffering from the same labors as those experienced during their oppression (Duvall, 2014). Hemp was not a profitable crop in the absence of free labor. Consequently, hemp farmers in Kentucky and other states that formerly had significant slave populations necessarily turned to other, less labor-intensive crops.

Hemp would see a brief U.S. rebound in the 1880s as new lands opened up on the Great Plains (Duvall, 2014). Increased grain production drove the demand for cordage and burlap to ship the commodities, and a binding machine that could tie cord was developed, increasing the need for strong fiber. Regional hemp production in the Great Plains duly followed, although this period was short-lived.

Markets for hemp at the turn of the 20th century faced long-term declines in demand as sailing ships gave way to steam and fossil-fuel powered vessels, thus reducing the need for rope and cordage (Fortenberry and Bennet, 2001, USDA, 2000). Cheaper imported fibers such as jute and abaca further cut into the hemp market (Dewey, 1914) and synthetic fibers loomed just around the corner. Still, at the turn of the century hemp was considered critical to the U.S. interest and USDA research supported its continued viability in the coming decades.

Hemp Fiber Production and Processing

As noted above, growing and processing hemp has at best been demanding and, at times, could be lethal. For centuries, the bulk of the work for producing hemp fibers was performed by hand. Such was the labor required that Venetians had called hemp "(the plant) of a hundred operations (processing steps)" (Schaefer, 1945, cited by Duvall, 2014). The lack of mechanization largely persisted until the early 20th century and placed hemp at a disadvantage relative to other fiber crops.

Multiple steps for cultivation and seeding were typical for hemp production (Herndon, 1963). In some cases, the plant's requirement for nutrients meant that "dunging" was needed if fields were not on river bottoms that received occasional replenishment by floods. Then there was the harvest and processing. Harvests historically occurred twice in a season, as male plants were removed around the time of pollen shed and

females after going to seed (Clarke and Merlin, 2013). At harvest, hemp was cut close to the plant crown or pulled up from the soil, arduous work even for the strongest men (Duvall, 2014). Cutting with a hemp knife (versus pulling up the roots) was a step forward when hemp made its way to Kentucky, and mechanized harvest was a novelty in the late 1800s (Dodge, 1896).

Following harvest, the plants had to be dried and retted before the long fibers could be extracted. Retting (a different way to say 'rotting') involves the microbially mediated decay of the bonds between the short inner fibers (hurd) and the long outer (bast) fibers in the hemp stem. The retting process removes lignin, pectin, waxes, and minor compounds and disaggregates "the pectin–lignin matrix that bounds the elementary hemp fibers and created fiber bundles" (Sisti et al., 2016).

Historically, the highest quality fibers have been obtained by water retting. This involves soaking hemp stalks in stagnant ponds of water and yields fibers with greater strength, lighter color, and greater consistency (Herndon, 1963) than obtained with other retting methods. Although the water retting was the typical method for processing hemp in the Baltics, the process was both tedious and unpleasant. French farmers considered water retting poisonous (Duvall, 2014), and U.S. producers seldom deployed it. The anaerobic conditions of the ponds created a putrid, noisome, and unhealthy environment for workers and surrounding communities. Producers in the United States wanted little to do with the process. In the 1840s, to improve the supply of naval-grade cordage, the Federal Government provided incentives for water retting but the added value of the fiber was not economical in the face of the lives of bondsmen lost to the endeavor, and the practice was abandoned (Bidwell, 1925).

A second method, summer retting, involved spreading the crop on the ground each evening to collect dew and bunching the material together the next day to keep the material moist. This practice was repeated until the fibers were suitably retted– but it was the method least employed by hemp growers given the high labor demands and the interference with concurrent duties associated with other crops (Herndon, 1963). More commonly, hemp was winter retted in the English colonies. At harvest, the crop was bundled into sheaves and placed along fences or left standing shocks. These later were spread on the ground after the warm season had ended, allowing the fibers to break down over the course of two or three winter months (Herndon, 1963).

Following retting, the stalks were stacked in shocks and dried again before decorticating. The decortication process, that is, the breaking the stalk's long outer bast fibers from the short interior pith or 'hurd', traditionally was done by hand with a hemp brake (Fig. 5). This required some amount of strength and skill to break the stalks without damaging the long fibers (Wright, 1918). The "broken" fibers then were "scotched" or scraped to remove the bulk of the hurd before the combing or "hackling'" process was used to straighten the bast fibers and remove residual impurities. Once fibers were satisfactorily extracted and straightened, they were spun into thread or yarn for woven products or cordage.

Clearly, hemp production was extremely physically demanding. The reasons why the industry so long remained reliant on outdated technologies and practices seems less apparent. Where mechanization had begun to supplant human labor and lower costs for other plant fibers, the hemp industry remained slow to change. As an early USDA report noted, "hemp is cleaned in the field, the cumbersome slat brake…in use for a hundred years or more in Kentucky being still employed" (Dodge, 1892).

Recognizing the need for mechanization, USDA supported research in Wisconsin and California (Fig. 6.) in the early 1900s. As one researcher mused, "no progress with hemp could be made as long as the crop was dependent on hand labor" (Wright, 1918). Application of engineering technologies to hemp production (assisted by declining European supplies following World War I) would help hemp to flourish in Wisconsin before political circumstances began to stifle the industry.

Hemp's Fall in the West

Early 20th century efforts of researchers such as Lyster Dewey at USDA and A.H. Wright at the University of Wisconsin would make significant progress for bringing hemp production into the modern age. Wright worked to develop cropping rotations, mechanize production practices, and support centralized processing centers (within regions suited for the crop) to economically supply Wisconsin processors (Wright, 1918). Dewey's breeding program created several hemp varieties and made substantial progress in stand yield (Dewey, 1928).

Despite these engineering and breeding advances, hemp's downward trajectory soon began due to conflation of the industrial crop with psychotropic strains of cannabis. Concern about the effects of recreational cannabis use came to the attention of the U.S. Federal Bureau of Narcotics and President Franklin D. Roosevelt, who supported legislation to restrict the production of psychoactive cannabis varieties (Ehrensing, 1998). In spite of opposition from the American Medical Association, the Marihuana Tax Act (MTA) passed in 1937, and it was from this point that the moniker "hemp" would begin to give way to marihuana/ marijuana for all things related to cannabis. The MTA placed cultivation of all *Cannabis* under control of the U.S. Treasury Department (USDA, 2000) and required growers to register and obtain licensure from the federal government. It was not an outright ban, but certainly a powerful effort to significantly reduce hemp production.

There is no small irony in the fact that in the subsequent year Popular Mechanics

FIG. 38.—Kentucky hemp brake.

Fig. 5. Image of a hemp brake for decorticating hemp. Image from Dodge (1897).

Fig. 6. Hemp production was tested in California around the turn of the 20th century. By the early 1900s California became the third largest hemp producer in the US, following Kentucky and Wisconsin. Image from the 1903 Yearbook of Agriculture (Dewey, 1904).

(Anonymous, 1938) published an article suggesting hemp would be the "new billion dollar crop." A billion dollars at that time was unimaginable to persons of almost any means, and the article declared "over 25,000 uses for the plant ranging from dynamite to cellophane." Hemp was coming into its own as a viable crop for North American farmers and a potential resource for literally thousands of consumer goods at just the wrong time.

Although some production persisted in Wisconsin, the constraints of the MTA effectively stifled the crop in the United States until fiber supplies to this country were interrupted by events during World War II. Several thousand farmers were thus recruited to grow "Hemp for Victory" (Johnson, 1999) and the USDA's Commodity Credit Corporation contracted War Hemp Industries, Inc. to construct several processing mills in the Midwest. Production peaked in 1943 to 1944 (USDA, 2000), only to decline again in the face of competition from cheaper imported fibers, the development of synthetic fibers, and renewed legal restrictions after the war.

Hemp in Eastern Europe and its Western Revival

Although numerous claims and conspiracy theories can be found concerning the reasons for hemp's demise in the west, these mostly are overblown (Duvall, 2014). In reality, hemp fibers faced the same challenges in Eastern Europe and Russia through the middle of the 20th century (Duvall, 2014). Demand for hemp declined in the face of cheaper alternatives and new synthetic filaments; new celluloid-based adhesive tapes also reduced the need for packaging twine. These issues, coupled with the restrictions surrounding hemp's potential for drug use were effective at suppressing the crop's production in the West.

Perhaps because of the long history of use or because of the fewer apprehensions about misuse as a drug (indeed, the low levels of THC may not have made this a concern) cultivation of hemp did not die in Eurasia. Breeding programs continued in attempts to develop uniform hybrids and monoecious (having both male and female flowers on the same plant) varieties. Much of this work, conducted from the 1930s through the 1960s, occurred in the former Soviet Union and Communist Block countries (ArynŠtejn and Hrennikova, 1967; Bócsa, 1958; Breslavec and Zaurov, 1937; Davidjan, 1963; Grecuhin and Belovickaja, 1940; Nevinnyh, 1962; Nikiforov, 1958; Rjazanskaja, 1963; Sizov, 1934). Hemp did not completely die in the west, either, as research continued in Italy from the 1930s to address issues related to agronomic practices and fiber quality (Zatta et al., 2012).

Although Western European countries had legal grounds for allowing hemp research in the 1970s (EC, 1971), efforts to revive industrial hemp as an agronomic crop in areas further to the west largely arose during the 1990s (Health Canada, 1998, EC, 1998). Production has been allowed in the European Union and in Canada provided hemp varieties contained less than 0.2% or 0.3% THC (Europe and Canada, respectively Since that time a raft of studies have evaluated a variety of items from agronomics to breeding to production and processing systems and end uses (e.g., see Fike, 2016).).

It is interesting to note here the modern and entirely artificial definition of hemp ($\leq$ 0.3% THC) versus marijuana (> 0.3% THC), as both are clearly members of the species, *Cannabis sativa*. As is obvious throughout this chapter, hemp has been the most common reference name of *Cannabis sativa* until the passage of the MTA. We often refer to the founding fathers of the United States (e.g., George Washington and Thomas Jefferson) as hemp farmers and even as vigorous supporters of hemp production. At that time, there was no understanding of the biochemistry of the species that now defines cannabinoids. We did not know what THC was; only that some hemp plants were intoxicating. Yet, we often refer to the cannabis grown during that era as hemp, even though it seems far more likely it would have produced levels of THC that exceeded our definition of marijuana today. Did Jefferson and others both promote and grow marijuana? Most would agree they most certainly did, based on the standards used to define these crops today. At this writing and in all but two developed countries (Uruguay and Canada), it is still wholly illegal grow marijuana, but in these same and other countries, it is also legal to grow hemp.

In the United States, grassroots lobbying efforts helped state governments to recognize undue restrictions on hemp. Although several states authorized feasibility studies to determine its potential value as a crop (USDA, 2000), restrictions imposed by the U.S. Drug Enforcement Agency (DEA) initially impeded this work. DEA's continued treatment of hemp as a Schedule I controlled substance, regardless of its THC content, made work with hemp as a crop all but impossible, particularly at a commodity scale.

The political winds buffeting hemp production have changed markedly in the 21st century. In the United States, hemp as a potential commodity crop largely has been rehabilitated with support from both ends of the political spectrum. Passage of the U.S. Farm Bill (signed into law as the Agriculture Act of 2014) created room for hemp study through section 7606, on "The Legitimacy of Industrial Hemp Research" (U.S. Congress, 2014). Several states now are actively engaged in hemp research and even vigorously supporting the development of hemp industries. Although answers to questions of "when" (rather than "if") hemp's outright legalization will occur remain unknown, these changes to the law bode well for hemp to legally return to U.S. production fields with the support of the federal government.

The Future for Hemp?

Much has been made of the potential for industrial hemp. Indeed, the crop's genetic potential to produce high quality seeds and fibers and the multitude of products (Fig. 7.) for which its constituents can be deployed certainly warrants the renewed exploration. Whether it can live up to claims of being "[h]umankind's savior" (!) (see Cherney and Small, 2016) undoubtedly is overblown and unlikely (and frankly, unnecessary and perhaps even counterproductive). The growing number of potential industrial applications for the plant give testament to the fact that hemp could exist once again as a valued member of the pantheon of agricultural crops. Conspiracy theorists that suggest "conniving industrialists and politicians... defeated hemp in the 1930s to favour competing industries...neglect much economic history" (Duvall, 2014). Ultimately, economics will again arbitrate the place for hemp, assuming growers, industry, and consumers have the ability to pursue such opportunities and uses to their full potential.

Summary

Cannabis is thought to have originated in central Asia. From the time of human interaction and intervention with the species, bands, clans, and tribes drove its dispersal across Asia and into Europe. The plant's seeds and fibers have played significant roles in meeting the basic needs of humans and it was an important plant material for ritual and religious experience. Cannabis played an important role in the advancement of human societies through its contributions to "string technologies" and likely contributed to the development of agriculture. Over time, Cannabis would become critical to European empires and nation-states that depended on its fibers as a means of obtaining wealth and power. From the 1500s, European nations carried the crop to the western hemisphere in efforts to expand supplies. Historically, hemp fiber systems were slow to mechanize and economically successful fiber production largely was based on captive (feudal or slave) labor. Use as a "war crop" also features prominently in hemp's history in the United States, with demand driven both by need and by lack of access to cheaper, imported fibers. Restrictions on

Industrial Hemp Seed and Stalk Processing and Products

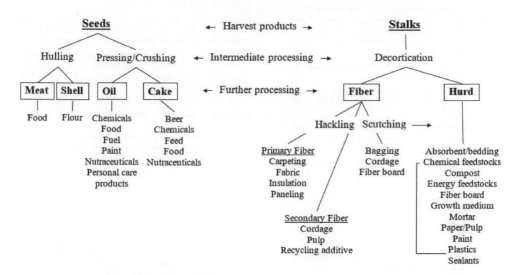

Fig. 7. Industrial hemp products and processing routes. Modified and adapted from Kraenzel et al. (1998). Decortication breaks the inner fibers from the long outer 'bast' fibers of the phloem. Scutching, a scraping process, removes the inner core fibers, or 'hurd'; hackling involves "brushing" and aligning the fibers prior to processing for textiles (Small, 2015).

hemp production occurred during the middle quarters of the 20th century in much of the West, put in place based on concerns over recreational drug use. Production and plant improvement efforts that continued with hemp during that time were largely attributed to the Soviet Union and countries of the Communist Bloc. Growing recognition of the distinction between fiber and recreational cannabis (and perhaps a changing of attitudes toward recreational use) have helped rehabilitate hemp as a commodity crop in the West. Hemp's primary use, historically, has been as a fiber crop and new process methods and uses are being found for the plant's fiber fractions. Production strictly as a grain crop is largely a 20th century construct. Growing recognition of the potential nutritional and health benefits of hemp seeds and essential oils from flowers contributes significantly to making this an important part of the hemp production portfolio. Hemp may have much to offer as a food, feed, fiber, fuel, and nutraceutical crop, but it remains to be seen how it will compete as a commodity crop. Ultimately, and provided that government interventions no longer hinders its use, economics will be the final arbiter of hemp's success as a commodity crop; a very simple matter of supply and demand.

References

Adovasio, J.M., O. Soffer, and J. Page. 2007. The invisible sex: Uncovering the true roles of women in prehistory. Routledge, New York.

Andresen, S.T., and S. Karg. 2011. Retting pits for textile fibre plants at Danish prehistoric sites dated between 800 B.C. and A.D. 1050. Veg. Hist. Archaeobot. 20:517–526. doi:10.1007/s00334-011-0324-0

Anonymous. 1938. New billion-dollar crop. Popular Mechanics February (238-239):144A–145A.

Aper, J., B. de Cauwer, S. de Roo, M. Lourenço, V. Fievez, R. Bulcke, and D. Reheul. 2014. Seed germination and viability of herbicide resistant and susceptible Chenopodium album populations after ensiling, digestion by cattle and manure storage. Weed Res. 54:169–177. doi:10.1111/wre.12063

Barber, E.J.W. 1992. Prehistoric textiles: The development of cloth in the Neolithic and Bronze Ages with special reference to the Aegean. Princeton Univ. Press, Princeton, NJ.

Benet, S. 1975. Early diffusion and folk uses of hemp. In: V. Rubin, editor, Cannabis and culture. Mouton & Co., The Hague. doi:10.1515/9783110812060.39

Bertinelli, L., and D. Black. 2004. Urbanization and growth. J. Urban Econ. 56:80–96. doi:10.1016/j.jue.2004.03.003

Bertoli, A., S. Tozzi, L. Pistelli, and L.G. Angelini. 2010. Fibre hemp inflorescences: From crop-residues to essential oil production. Ind. Crops Prod. 32:329–337. doi:10.1016/j.indcrop.2010.05.012

Bidwell, P.W. 1925. History of agriculture in the northern United States, 1620-1860. Carnegie Institution of Washington. The University of Chicago Press. Chicago, IL.

Bócsa, I., Z. Finta-Korpelová, and P. Máthé. 2005. Preliminary results of selection for seed oil content in hemp (Cannabis sativa L.). J. Ind. Hemp 10:5–15. doi:10.1300/J237v10n01_02

Bremmer, J.N. 2002. The rise and fall of the afterlife: The 1995 Read-Tuckwell Lectures at the University of Bristol. Routlege, London.

Brombacher, C., and D. Hecker. 2015. Agriculture, food and environment during Merovingian times: Plant remains from three early medieval sites in northwestern Switzerland. Veg. Hist. Archaeobot. 24:331–342. doi:10.1007/s00334-014-0460-4

Callaway, J.C. 2003. Variety: 'Finola'. Application no. 2001/003. Plant Var. J. 16:30–31.

Callaway, J.C. 2004. Hempseed as a nutritional resource: An overview. Euphytica 140:65–72. doi:10.1007/s10681-004-4811-6

Campos, I. 2012. Home grown marijuana and the origins of Mexico's war on drugs. University of North Carolina Press, Chapel Hill, NC.

Chang, K.C. 1968. The archeology of ancient China. 2nd ed. Yale University. London. doi:10.1126/science.162.3853.519

Chen, W., W. Wang, and X. Dai. 2009. Holocene vegetation history with implications of human impact in the Lake Chaohu area, Anhui Province, East China. Veg. Hist. Archaeobot. 18:137–146. doi:10.1007/s00334-008-0173-7

Cheng, T.K. 1982. Studies in Chinese Archaeology. Chinese Univ. Press, Hong Kong.

Cherney, H.J., and E. Small. 2016. Industrial hemp in North America: Production, politics and potential. Agronomy (Basel, Switz.) 6. doi:10.3390/agronomy6040058

Clarke, R., and M. Merlin. 2013. Cannabis: Evolution and ethnobotany. University of California Press, Berkeley, CA. (Accessed October 2017).

Clarke, R.C. 2002. The history of hemp in Norway. J. Ind. Hemp 7:89–103. doi:10.1300/J237v07n01_08

Coetzee, C., R.A. Levendal, M. van de Venter, and C.L. Frost. 2007. Anticoagulant effects of a cannabis extract in an obese rat model. Phytomedicine 14:333–337. doi:10.1016/j.phymed.2006.02.004

Cozma, A., S. Andrei, A. Pintea, D. Miere, L. Filip, F. Loghin, and A. Ferlay. 2015. Effect of hemp seed oil supplementation on plasma lipid profile, liver function, milk fatty acid, cholesterol, and vitamin A concentrations in Carpathian goats. Czech J. Anim. Sci. 60:289–301. doi:10.17221/8275-CJAS

Cyprien, A.L., L. Visset, and N. Carcaue. 2004. Evolution of vegetation landscapes during the Holocene in the central and downstream Loire basin (Western France). Veg. Hist. Archaeobot. 13:181–196. doi:10.1007/s00334-004-0042-y

Darwin, C.R. 1869. Origin of species. Online Variorum of Darwin's Origin of Species. Fifth British ed. Darwin Online. http://darwin-online.org.uk/ (Accessed 5 Mar. 2019).

Devane, W.A., F.A. Dysarz, III, M.R. Johnson, L.S. Melvin, and A.C. Howlett. 1988. Determination and characterization of a cannabinoid receptor in rat brain. Mol. Pharmacol. 34:605–613.

Dewey, L.H. 1904. Principle commercial plant fibers. In: USDA, editor, Yearbook of the United States Department of Agriculture–1903. Government Printing Office, Washington, D.C.

Dewey, L.H. 1914. Hemp. Yearbook of the United States Department of Agriculture– 1913. Government Printing Office, Washington, D.C. p. 283–346.

Dewey, L.H. 1928. Hemp varieties of improved type are result of selection In: USDA, editor, What's new in agriculture. Yearbook of the United States Department of Agriculture–1927. Government Printing Office, Washington. p. 357–360.

Dodge, C.R. 1892. A report on flax, hemp, ramie, and jute, with consideration upon flax and hemp culture in Europe, a report on teh ramie machine trials of 1889 in Paris, and present status of fiber industries in the United States. Fiber investigations Report No. 1. USDA, Washington, D.C.

Dodge, C.R. 1896. Hemp culture. In: USDA, editor. Yearbook of the United States Department of Agriculture–1895. Government Printing Office, Washington, D.C.

Dodge, C.R. 1897. A descriptive catalogue of useful fiber plants of the world, including the structural and ecnomic classifications of fibers. Report No. 9. USDA Office of Fiber Investigations, Washington, D.C.

Duvall, C.S. 2014. Cannabis. Reaktion Books, London, U.K.

European Commission (EC). 1971. Regulation (EEC) No 619/71 of the council of 22 March 1971 laying down general rules for granting aid for flax and hemp. Euopean Commision. Official Journal of the European Communities, Brussels. p. 169-170.

European Commission (EC). 1998. Council Regulation (EC) No 1420/98 of 26 June 1998 amending Regulation (EEC) No 619/71 laying down general rules for granting aid for flax and hemp. European Commission. Official Journal of the European Communities, Luxembourg. p. L 190/197-198.

European Food Safety Authority (EFSA). 2015. Scientific opinion on the risks for human health related to the presence of tetrahydrocannabinol (THC) in milk and other food of animal origin. EFSA J. 13:4141.

European Food Safety Authority (EFSA). 2011. Scientific Opinion on the safety of hemp (Cannabis genus) for use as animal feed. EFSA J. 9:1–41.

Ehrensing, D.T. 1998. Feasibility of industrial hemp production in the United States Pacific Northwest. Oregon State University Extension Service, Corvallis, OR.

Fægri, K., P.E. Kaland, and K. Krzywinski. 1989. Textbook of pollen analysis. John Wiley & Sons Ltd., Chichester, U.K.

Fernández-Ruiz, J., O. Sagredo, M.R. Pazos, C. García, R. Pertwee, R. Mechoulam, and J. Martínez-Orgado. 2013. Cannabidiol for neurodegenerative disorders: Important new clinical applications for this phytocannabinoid? Br. J. Clin. Pharmacol. 75:323–333. doi:10.1111/j.1365-2125.2012.04341.x

Fike, J. 2016. Industrial hemp: Renewed opportunities for an ancient crop. Crit. Rev. Plant Sci. 35:406–424. doi:10.1080/07352689.2016.1257842

Fleming, M.P., and R.C. Clarke. 1998. Physical evidence for the antiquity of Cannabis sativa L. Journal of the International Hemp Association 5:80–93.

Fortenberry, T.R., and M. Bennet. 2001. Is industrial hemp worth further study in the U.S.? A survey of the literature. Wisconsin Madison Agricultural & Applied Economics Staff Papers, University of Wisconsin-Madison, Madison, WI.

Gepts, P. 2004. Crop domestication as a long-term selection experiment. Plant Breed. Rev. 24:1–44.

Gibb, D.J., M.A. Shah, P.S. Mir, and T.A. McAllister. 2005. Effect of full-fat hemp seed on performance and tissue fatty acids of feedlot cattle. Can. J. Anim. Sci. 85:223–230. doi:10.4141/A04-078

Girgih, A.T., A. Alashi, R. He, S. Malomo, and R.E. Aluko. 2014a. Preventive and treatment effects of a hemp seed (Cannabis sativa L.) meal protein hydrolysate against high blood pressure in spontaneously hypertensive rats. Eur. J. Nutr. 53:1237–1246. doi:10.1007/s00394-013-0625-4

Girgih, A.T., A.M. Alashi, R. He, S.A. Malomo, P. Raj, T. Netticadan, and R.E. Aluko. 2014b. A novel hemp seed meal protein hydrolysate reduces oxidative stress factors in spontaneously hypertensive rats. Nutrients 6:5652–5666. doi:10.3390/nu6125652

Górski, R., M. Szklarz, and R. Kaniewski. 2009. Efficacy of hemp essential oil in the control of rosy apple aphid (Dysaphis plantaginea Pass.) occurring on apple tree. Prog. in Plant Prot. 49:2013–2016.

Gray, L.C., and E.K. Thompson. 1933. History of agriculture in the southern United States to 1860. Vol. 1. Carnegie Institution of Washington. The University of Chicago Press, Chicago, IL.

Hampson, B.A., M.A. de Laat, P.C. Mills, and C.C. Pollitt. 2010. Distances travelled by feral horses in 'outback' Australia. Equine Vet. J. 42:582–586. doi:10.1111/j.2042-3306.2010.00203.x

Harlan, J.R. 1965. The possible role of weed races in the evolution of cultivated plants. Euphytica 14:173–176. doi:10.1007/BF00038984

Health Canada. 1998. Hemp and the hemp industry: Frequently Asked Questions. Health Canada, Ottawa, Canada. https://www.canada.ca/en/health-canada/services/drugs-medication/cannabis/producing-selling-hemp/about-hemp-canada-hemp-industry/frequently-asked-questions.html#a7 (Accessed 1 Feb. 2019).

Herndon, G.M. 1963. Hemp in colonial Virginia. Agr. Hist. 37:86–93.

Herndon, G.M. 1966. A war-inspired industry. Va. Mag. Hist. Biogr. 74:301–311.

Hessle, A., M. Eriksson, E. Nadeau, T. Turner, and B. Johansson. 2008. Cold-pressed hempseed cake as a protein feed for growing cattle. Acta Agriculturæ Scandinavica. Section A. Anim. Sci. 58:136–145. 10.1080/09064700802452192

Hillig, K.W., and P.G. Mahlberg. 2004. A chemotaxonomic analysis of cannabinoid variation in Cannabis (Cannabaceae). Am. J. Bot. 91:966–975. doi:10.3732/ajb.91.6.966

Husbands, J.D. 1909. Seeds and plants imported during the period from October 1 to December 31, 1908. In: B.T. Galloway, editor, USDA Bureau of Plant Industry Bulletin, Washington, D.C. p. 42.

Ignatov, A.N., A.M. Artemyeva, and K. Hida. 2010. Origin and expansion of cultivated Brassica rapa in Eurasia: Linguistic facts. Acta Hortic. 867:81–88. doi:10.17660/ActaHortic.2010.867.9

Janischewsky, D.E. 1924. Forma konopli na sornykh mestakh v Yugovostochnoi Rossii (The form of cannabis in weedy places in southeast Russia) In: I.A. Chiuevsky, editor, Uchenye zapiski gosudarstvennogo saratovskogo imeni N.G. chernyshevskogo universiteta. Saratov Univ. Press, Saratov, USSR. 2:3–17.

Jia, P.W.M. 2007. Transition from foraging to farming in northeast China. PhD thesis. University of Sydney, Sydney, Australia.

Johnson, P. 1999. Industrial hemp: A critical review of claimed potentials for Cannabis sativa. Tappi J. 82:113–123.

Karlsson, L., M. Finell, and K. Martinsson. 2010. Effects of increasing amounts of hempseed cake in the diet of dairy cows on the production and composition of milk. Animal 4:1854–1860. doi:10.1017/S1751731110001254

Kaul, N., R. Kreml, J.A. Austria, M.N. Richard, A.L. Edel, E. Dibrov, S. Hirono, M.E. Zettler, and G.M. Pierce. 2008. A comparison of fish oil, flaxseed oil and hempseed oil supplementation on selected parameters of cardiovascular health in healthy volunteers. J. Am. Coll. Nutr. 27:51–58. doi:10.1080/07315724.2008.10719674

Keng, H. 1973. Economic plants of ancient North China as mentioned in Shih Ching (Book of Poetry). Econ. Bot. 28:391–410. doi:10.1007/BF02862856

Kraenzel, D.G., T. Petry, B. Nelson, M.J. Anderson, D. Mathern, and R. Todd. 1998. Industrial hemp as an alternative crop in North Dakota. Agricultural Economics Report. North Dakota State University, Fargo, ND. p. 22.

Kriese, U., E. Schumann, W.E. Weber, M. Beyer, L. Brühl, and B. Matthäus. 2004. Oil content, tocopherol composition and fatty acid patterns of the seeds of 51 Cannabis sativa L. genotypes. Euphytica 137:339–351. doi:10.1023/B:EUPH.0000040473.23941.76

Kudo, Y., M. Kobayashi, A. Momohara, S. Noshiro, T. Nakamura, S. Okitsu, S. Yanagisawa, and T. Okamoto. 2009. [Japanese with Abstract in English] Radiocarbon dating of fossil hemp fruits in the earliest Jomon period excavated from the Okinoshima site, Chiba, Japan. Japanese Journal of Historical Botany 17:27–31.

Lee, G.-A., G.W. Crawford, L. Liu, and X. Chen. 2007. Plants and people from the Early Neolithic to Shang periods in North China. Proc. Natl. Acad. Sci. USA 104:1087–1092. doi:10.1073/pnas.0609763104

Leson, G. 2013. Hemp seeds for nutrition. CABI, Wallingford. p. 229-238.

Li, H.L. 1973. An archaeological and historical account of cannabis in China. Econ. Bot. 28:437–448. doi:10.1007/BF02862859

Li, H.L. 1974. The origin and use of cannabis in Eastern Asia. Linguistic-cultural implications. Econ. Bot. 28:293–301. doi:10.1007/BF02861426

Li, M., X. Yang, H. Wang, Q. Wang, X. Jia, and Q. Ge. 2010. Starch grains from dental calculus reveal ancient plant foodstuffs at Chenqimogou site, Gansu Province. Sci. China Earth Sci. 53:694–699. doi:10.1007/s11430-010-0052-9

Li, X., X. Shang, J. Dodson, and X. Zhou. 2009. Holocene agriculture in the Guanzhong Basin in NW China indicated by pollen and charcoal evidence. Holocene 19:1213–1220. doi:10.1177/0959683609345083

Liu, F., H. Hu, G. Du, G. Deng, and Y. Yang. 2017. Ethnobotanical research on origin, cultivation, distribution and utilization of hemp (Cannabis sativa L.) in China. Indian J. Tradit. Knowl. 16:235–242.

Long, T., M. Wagner, D. Demske, C. Leipe, and P.E. Tarasov. 2017. Cannabis in Eurasia: Origin of human use and Bronze Age trans-continental connections. Veg. Hist. Archaeobot. 26:245–258. doi:10.1007/s00334-016-0579-6

Meijer, E.P.M.d., M. Bagatta, A. Carboni, P. Crucitti, V.M.C. Moliterni, P. Ranalli, and G. Mandolino. 2003. The inheritance of chemical phenotype in Cannabis sativa L. Genetics 163:335–346.

Mercuri, A.M., C.A. Accorsi, and M.B. Mazzanti. 2002. The long history of Cannabis and its cultivation by the Romans in central Italy, shown by pollen records from Lago Albano and Lago di Nemi. Veg. Hist. Archaeobot. 11:263–276. doi:10.1007/s003340200039

Molodkov, A., and N. Bolikhovskaya. 2006. Long-term palaeoenvironmental changes recorded in palynologically studied loess–palaeosol and ESR-dated marine deposits of Northern Eurasia: Implications for sea-land correlation. Quat. Int. 152–153:37–47. doi:10.1016/j.quaint.2005.12.010

Mourot, J., and M. Guillevic. 2015. Effect of introducing hemp oil into feed on the nutritional quality of pig meat. OCL- Oilseeds and Fats. Crops and Lipids 22:D612.

Mukherjee, A., S.C. Roy, S. de Bera, H. Jiang, X. Li, C. Li, and S. Bera. 2008. Results of molecular analysis of an archaeological hemp (Cannabis sativa L.) DNA sample from North West China. Genet. Resour. Crop Evol. 55:481–485. 10.1007/s10722-008-9343-9

Nishida, T., N. Shimizu, M. Ishida, T. Onoue, and N. Harashima. 1998. Effect of cattle digestion and of composting heat on weed seeds. JARQ. Jpn. Agric. Res. Q. 32:55–60.

Novak, J., K. Zitterl-Eglseer, S.G. Deans, and C.M. Franz. 2001. Essential oils of different cultivars of Cannabis sativa L. and their antimicrobial activity. Flavour Fragrance J. 16:259–262. doi:10.1002/ffj.993

Okazaki, H., M. Kobayashi, A. Momohara, S. Eguchi, T. Okamoto, S. Yanagisawa, S. Okubu, and J. Kiyonaga. 2011. Early Holocene coastal environment change inferred from deposits at Okinoshima archeological site, Boso Peninsula, central Japan. Quaternary International 230:87–94. doi: 10.1016/j.quaint.2009.11.002.

Oomah, B.D., M. Busson, D.V. Godfrey, and J.C.G. Drover. 2002. Characteristics of hemp (Cannabis sativa L.) seed oil. Food Chem. 76:33–43. doi:10.1016/S0308-8146(01)00245-X

Harper, Douglas. 2019. Online etymology dictionary. Etym Online. https://www.etymonline.com/ (Accessed 1 Feb. 2019). [2019 is year accessed].

Parian, A., and B.N. Limketkai. 2016. Dietary supplement therapies for inflammatory bowel disease: Crohn's disease and ulcerative colitis. Curr. Pharm. Des. 22:180–188. doi:10.2174/1381612822666151112145033

Pathak, T., J. Kaur, R. Kumar, and K. Kuldeep. 2016. Development of electrochemical biosensor for detection of asparagine in leukemic samples. Int. J. Pharm. Sci. Res. 7:783–788 (IJPSR).

Pope Innocent VIII. 1484. Summis desiderantes affectibus (Papal Bill of 1484). Medieval European History Archives, University of Montevallo, Montevallo, AL. http://carmichaeldigitalprojects.org/hist447/items/show/67 (Accessed 1 Feb. 2019).

Poska, A., and L. Saarse. 2006. New evidence of possible crop introduction to north-eastern Europe during the Stone Age. Veg. Hist. Archaeobot. 15:169–179. doi:10.1007/s00334-005-0024-8

Pringle, H. 1997. Ice age communities may be earliest known net hunters. Science 277:1203–1204. doi:10.1126/science.277.5330.1203

Prociuk, M.A., A.L. Edel, M.N. Richard, N.T. Gavel, B.P. Ander, C.M.C. Dupasquier, and G.N. Pierce. 2008. Cholesterol-induced stimulation of platelet aggregation is prevented by a hempseed-enriched diet. 86:153-159. doi:10.1139/y08-011.

Quinn, L.D., M. Kolipinski, V.R. Coelho, B. Davis, J.M. Vianney, O. Batjargal, M. Alas, and S. Ghosh. 2008. Germination of invasive plant seeds after digestion by horses in California. Nat. Areas J. 28:356–362. doi:10.3375/0885-8608(2008)28[356:GOIPSA]2.0.CO;2

Radu, S., and T. Robu. 2014. Effects and efficiency of dietary hemp seed and flaxseed oils on the human metabolic function. J. Environ. Prot. Ecol. 15:326–331.

Rahimi, S., H.R. Mashhadi, M.D. Banadaky, and M.B. Mesgaran. 2016. Variation in weed seed fate fed to different Holstein cattle groups. PLoS One 11:E0154057. 10.1371/journal.pone.0154057

Rahn, B., B.J. Pearson, R.N. Trigiano, and D.J. Gray. 2016. The derivation of modern cannabis varieties. Crit. Rev. Plant Sci. 35:328–348. doi:10.1080/07352689.2016.1273626

Rieder, S.A., A. Chauhan, U. Singh, M. Nagarkatti, and P. Nagarkatti. 2010. Cannabinoid-induced apoptosis in immune cells as a pathway to immunosuppression. Immunobiology 215:598–605. doi:10.1016/j.imbio.2009.04.001

Rosetti, D.V. 1959. Movilele funerare de la Gurb ne ti In: Academiei Republicii Populare Romîne, VI, editor, X. Materiale i cercetari arheologice. (In Romanian). Academia Republicee Populare Romine, Bucharest, Romania. http://www.samorini.it/doc1/alt_aut/lr/rosetti-movilele-funerare-de-la-gurbanesti.pdf (Accessed 1 Feb. 2019).

Roulac, J.W. 1997. Hemp horizons: The comeback of the world's most promising plant. Chelsea Green Publishing Company, White River Junction, VT.

Russo, E. 2004. History of cannabis as a medicine. In: G. Guy, B. Whittle, and P. Robson, editors, Medicinal uses of cannabis and cannabinoids. Pharmaceutical Press, London.

Sawler, J., J.M. Stout, K.M. Gardner, D. Hudson, J. Vidmar, L. Butler, J.E. Page, and S. Mylers. 2015. The genetic structure of marijuana and hemp. PLoS One 10:E0133292. doi:10.1371/journal.pone.0133292

Schaefer, G. 1945. The cultivation and preparation of flax and hemp. CIBA Review 49:1779–1795.

Schultes, R.E. 1970. Random thoughts and queries on the botany of Cannabis. In: R.B. Joyce and S.H. Curry, editors, The botany and chemistry of Cannabis. J. & A. Churchill, London. p. 11–38.

Schultes, R.E., and A. Hofmann. 1992. Plants of the gods: Their sacred, healing, and hallucinogenic powers. Healing Arts, Rochester, VT.

Schwab, U.S., J.C. Callaway, A.T. Erkkilä, J. Gynther, M.I.J. Uusitupa, and T. Järvinen. 2006. Effects of hempseed and flaxseed oils on the profile of serum lipids, serum total and lipoprotein lipid concentrations and haemostatic factors. Eur. J. Nutr. 45:470–477. doi:10.1007/s00394-006-0621-z

Sherratt, A.G. 1981. Plough and pastoralism: Aspects of the secondary products revolution. In: I. Hodder, G. Isaac, and N. Hammond, editors, Pattern of the past: Studies in honour of David Clarke. Cambridge University, Cambridge. p. 261–306.

Small, E. 2015. Evolution and classification of Cannabis sativa (marijuana, hemp) in relation to human utilization. Bot. Rev. 81:189–294.

Small, E. 2017. Classification of Cannabis sativa L. in relation to agricultural, biotechnological, medical and recreational utilization. In: S. Chandra, H. Lata, and M.A. ElSohly, editors, Cannabis sativa L.- Botany and Biotechnology. Springer International Publishing, Cham. p. 1–62. doi:10.1007/978-3-319-54564-6_1

Small, E., and D. Marcus. 2002. Hemp: A new crop with new uses for North America. ASHS Press, Alexandria. p. 284-326.

Small, E., D. Marcus, G. Butler, and A.R. McElroy. 2007. Apparent increase in biomass and seed productivity in hemp (Cannabis sativa) resulting from branch proliferation caused by the European corn borer (Ostrinia nubilalis). J. Ind. Hemp 12:15–26. doi:10.1300/J237v12n01_03

Sytsma, K.J., J. Morawetz, J.C. Pires, M. Nepokroeff, E. Conty, M. Zjhra, J.C. Hall, and M.W. Chase. 2002. Urticalean rosids: Circumscription, rosid ancestry, and phylogenetics based on rbcL, trnL-F, and ndhF sequences. Am. J. Bot. 89:1531–1546. doi:10.3732/ajb.89.9.1531

Thomas, T.G., S.K. Sharma, P. Anand, and B.R. Sharma. 2000. Insecticidal properties of essential oil of Cannabis sativa Linn. against mosquito larvae. Entomon 25:21–24.

Turner, T., A. Hessle, K. Lundström, and J. Pickova. 2008. Influence of hempseed cake and soybean meal on lipid fractions in bovine M. longissimus dorsi. Acta Agriculturæ Scandinavica. Section A. Anim. Sci. 58:152–160. 10.1080/09064700802492354

U.S. Congress. 2014. H.R. 2642. Agricultural act of 2014 (PL 113-79), 128 United States Statutes at Large. Government Printing Office, Washington, D.C.

USDA. 2000. Industrial hemp in the United States: Status and market potential USDA Economic Research Service, Washington, D.C. p. 43.

Vavilov, N.I. 1951. The origin, variation, immunity and breeding of cultivated plants. Chronica Botanica, Waltham, MA.

Vera, G., V. López-Miranda, E. Herradón, M.I. Martín, and R. Abalo. 2012. Characterization of cannabinoid-induced relief of neuropathic pain in rat models of type 1 and type 2 diabetes. Pharmacol. Biochem. Behav. 102:335–343. doi:10.1016/j.pbb.2012.05.008

Vonapartis, E., M.P. Aubin, P. Seguin, A.F. Mustafa, and J.B. Charron. 2015. Seed composition of ten industrial hemp cultivars approved for production in Canada. J. Food Compos. Anal. 39:8–12. doi:10.1016/j.jfca.2014.11.004

Warf, B. 2014. High points: A historical geography of Cannabis. Geogr. Rev. 104:414–438. doi:10.1111/j.1931-0846.2014.12038.x

Whitfield, S. 1999. Life along the Silk Road/Susan Whitfield. John Murray, London.

Woods, V.B., and A.M. Fearon. 2009. Dietary sources of unsaturated fatty acids for animals and their transfer into meat, milk and eggs: A review. Livest. Sci. 126:1–20. doi:10.1016/j.livsci.2009.07.002

Woods, V.B., and E.G.A. Forbes. 2007. Dietary sources of unsaturated fatty acids for animals and their availability in milk, meat and eggs: A summary of research findings. British Grassland Society Occasional Symposium No.38. British Grassland Society (BGS), Reading, U.K. p. 337–340.

Wright, A.H. 1918. Wisconsin's hemp industry. Bulletin 293. Agricultural Experiment Station of the Univeristy of Wisconsin. Madison, WI. p. 47.

Yang, M., R.v. Velzen, F.T. Bakker, A. Sattarian, D. Li, and T. Yi. 2013. Molecular phylogenetics and character evolution of Cannabaceae. Taxon 62:473–485. doi:10.12705/623.9

Yang, X., Z. Wan, L. Perry, H. Lu, Q. Wang, C. Zhao, J. Li, F. Xie, J. Yu, T. Cui, T. Wang, M. Li, and Q. Ge. 2012. Early millet use in northern China. Proc. Natl. Acad. Sci. USA 109:3726–3730. doi:10.1073/pnas.1115430109

Zatta, A., A. Monti, and G. Venturi. 2012. Eighty years of studies on industrial hemp in the Po Valley (1930–2010). J. Nat. Fibers 9:180–196. 10.1080/15440478.2012.706439

Zheng, X., L.D. Martin, Z. Zhou, D.A. Burnham, F. Zhang, and D. Miao. 2011. Fossil evidence of avian crops from the Early Cretaceous of China. Proc. Natl. Acad. Sci. USA 108:15904–15907. doi:10.1073/pnas.1112694108

Chapter 2: Hemp Grain

D.W. Williams, Ph.D.

Introduction

Hemp grain, as is the case for nearly all grain crops, can be harvested and utilized in multiple ways. Global production of hemp grain has fluctuated greatly over time according to consumer demand, and possibly also due to a lack of scientific research regarding optimal utilization of the different potential products. In this chapter, we will provide an overview of modern utilization of hemp grain in different realms; more specifically as food, personal care, or fuel products.

We begin by noting as did Small (2015), that from a botanical perspective, hemp produces achenes, which are fruits that contain seeds within. However, hemp grain is almost universally referred to as hemp seed. This is true at nearly all levels; from the United Nations reporting on global hempseed production, to farmers speaking with peers in the U.S. and other countries. As such, while botanically incorrect, we will use the terms hemp grain and hemp seed interchangeably in this work.

Hemp grain may be utilized as food for either humans or animals, just as are other common grains (e.g., corn, wheat, and others). The seeds or their derivatives may be used in different forms ranging from toasted whole seeds to processed oil that's been pressed or otherwise extracted from seeds. The physicochemical properties of hemp seeds and derivatives can be positive or negative from both digestive and nutritional perspectives. While not a perfect food in any form, products either containing or derived from hemp seeds can be both desirable and very marketable in today's society.

In addition to food, derivatives from hemp grain are thought to be potentially useful ingredients in personal care products. Within the modern hemp space, these products are often but not always referred to as cosmeceuticals, or cosmetic products that also provide some health benefit similar to that provided by a drug. An example may be a lotion containing hemp seed oil, the use of which could not only moisturize the skin at the application site, but theoretically could also reduce oxidative stress in cells at the application site by the antioxidants contained in the hemp seed oil. There are other purported potential effects from hemp-derived ingredients as well, but few if any that are supported by clinical evaluations.

At the same time, it is necessary to note that the U.S. Food and Drug Administration (FDA) does not recognize the term cosmeceutical, and further opines that the

---

University of Kentucky College of Agriculture, Lexington, KY. *Corresponding author (dwilliams@uky.edu)

doi:10.2134/industrialhemp.c2

Industrial Hemp as a Modern Commodity Crop. D.W. Williams, editor.

word has no meaning under the law (https://www.fda.gov/Cosmetics/GuidanceRegulation/LawsRegulations/ucm074201.htm, Accessed 5 Mar. 2019). The FDA regulates and approves all legal drug products, that is, "articles intended for use in the diagnosis, cure, mitigation, treatment, or prevention of disease" and "articles (other than food) intended to affect the structure or any function of the body of man or other animals". Conversely, the FDA does not review or approve cosmetic products for sale to or use by consumers.

Another potential use of hemp seeds and/or their derivatives could be as a source of bioenergy products. Hemp seed oils have the potential to produce high-quality biodiesel fuel (Li et al., 2010) or contribute to the production of bioethanol (Das et al., 2017).

It is very clear that interest in hemp seed utilization has increased significantly since the United States began work with the industrial hemp pilot research programs in 2014. A quick perusal of the current refereed scientific literature from 2014 to 2018 indicates no less than 65 refereed articles reporting on the utilization of hemp seeds, mostly as foods, but also as potential sources of bioenergy. Of those 65, over one-third (24) were published in 2018.

Generally speaking, and as is nearly always the case with natural products, hemp seed is not perfect in all regards. There are caveats and in some cases pure exceptions regarding optimal utilization in nearly any application. That said, hemp seeds do have real potential for modern utilization in all three realms; food, personal care products, and fuel. What must be determined is how hemp seed will compete with existing sources in each realm from both efficacious and cost-effective perspectives; the determination of which will ultimately define consumer demand.

Utilization as Food

Callaway (2004) provided an excellent overview of the utilization of hemp seeds as food. In that work, Callaway cited other workers as providing evidence of humans utilizing hemp grain as food in pre-historic times and potentially even by pre-human hominids. Regarding modern times, Callaway reported that hemp seed foods had not been broadly introduced into western markets in 2004.

Today, hemp grain production is distributed mostly in Canada, a few countries in the European Union (EU), and in China. According to Health Canada (2019), 138,018 acres were dedicated to hemp grain production in the 2017 crop year. This was distributed across 21 approved varieties with only 15 acres dedicated to unknown/unnamed germplasm, mostly used for plant breeding purposes. Major production areas included Alberta, Manitoba, Quebec, and Saskatchewan. Apparently, yield data from the 2017 crop year in Canada is not publicly available at time of this writing. European hemp grain production in 2017 was reported at 48,131 harvested acres yielding 94,513.16 tons of grain (FAO, 2019). China was reported to have harvested 16,424 tons from 14,564 acres in 2017 (FAO, 2019).

While yield data from Canada is not entire, Canadian farmers are clearly one of the largest producers and exporters of hemp grain for food. Evaluation of the production and yield data that is available (Johnson, 2018) indicates great variation in acreage planted during 2001 through 2015. Many have postulated this variation in production efforts is a direct reflection of the classical premise of supply and demand. For example, Johnson (2018) reported that Canadian acreage planted to hemp grain in 2007 was greatly reduced relative to 2005 to 2006 as a correction to overproduction, but also partially due to positive economics of other crops. If this premise is accurate, then a little less than 50,000 acres of production exceeded the demand for the crop at that time.

Put simply, we cannot know what the demand for food products that contain or are derived from hemp grain will ultimately be. Based on fluctuations in Canadian production over time, we can postulate that the current demand is being met and sometimes exceeded by existing production efforts. But, we must also recognize that relatively speaking, hemp food products are essentially new, especially products domestically produced within the United States. Hence, these products have not been marketed aggressively. It is likely a very safe assumption that appropriate and aggressive marketing will have a positive impact on consumer demand. However, we still cannot know what the ultimate total demand might become, or even just total demand within the U.S. alone.

One cannot help but consider consumer demands in the western cultures of Canada and the European Union since hemp has been a legal crop in those countries. While these cultures are clearly not identical to that of the United States, they are more similar than Asian, including Russian cultures. If we consider that hemp has been a legal crop and that foods containing or derived from hemp grain have been legal for 20+ years in Canada and the EU, why is the consumer demand in these western cultures so small if the benefits are so large? Why aren't more consumers in western cultures demanding hemp grain foods? Is it because they have not proved to be as tasteful and/or beneficial to our health as proposed?

Perhaps we could also consider an anecdotal evaluation of the perceived benefits derived from including hemp seeds in the human diet. Chikezie and Ojiako (2015) reported ancient China was a source of information on the use of medicinal plants and foods via the 'The Pen Tsao', published about 1600 BC. Gale and Null (2018) opined that holistic and herbal medicine has been effective in Chinese culture for millennia, continues to be effective today, and is even enjoying new investigative interest by many countries across the globe. If we accept that Asian, and more specifically Chinese culture, has provided many of the world's most impactful benefits from holistic and herbal medicine, and, hemp grain is exceptionally beneficial as a human food, why isn't hemp grain a much larger part of Asian diets today?

Hemp is native to China and other parts of Asia. There are areas within China where hemp production has never been regulated; even today. If there were direct health benefits from including hemp grain in our daily diets, how could the Chinese people have possibly overlooked this benefit? As mentioned above, total hemp grain production on the Chinese mainland in 2017 as reported by FAO was 16, 424 tons. Considering a human population of 1.386 billion that year, this would equate to about 11 g of grain (about 0.02 pound) per person during all of 2017. And, this assumes that none of the hemp grain produced in China was fed to animals or exported, which is certainly not accurate. Clearly, hemp grain is not a major contributor to the average diet of the Chinese people. Why not? Why don't the peoples living at the epicenter of impactful holistic and herbal medicine place more value on hemp grain as a food and source of health benefits? Is it even possible that they have overlooked the benefits of this super-food for the entirety of their existence on the mainland of China?

If we consider only modern rhetoric, perhaps we must consider that it is indeed possible. But, as noted above, over 65 refereed articles have been published 2014 to 2018, which is over 10 articles per year regarding hemp seed, mostly as an efficient source of nutritional and other health benefits. Can all of these reports be wrong or exaggerated? Almost certainly not. It seems rather apparent there are real potential benefits provided by hemp seeds and derivatives as foods as is purported by the recent, refereed scientific literature.

We will again refer to Callaway (2004) and Small (2015) for full consideration and evaluation of the direct physicochemical attributes of hemp seed and its derivatives as human and animal food. Here, we will concentrate mostly but not entirely on refereed reports generated after the work of Small (2015). Generally speaking, hemp seed and/or derivatives are reported to contain high protein content that is highly digestible, high oil content with positive fatty acid profiles including an approximate 3:1 ratio of omega 6 to omega 3 fatty acids, and high concentrations of antioxidant molecules.

An excellent evaluation of the nutritional attributes of hemp seed from several accessions (accessions are unnamed experimental lines that have been collected from endemic or natural populations) compared with improved cultivars was provided by Galasso et al. (2016). They reported many of the same positive attributes listed above, but also reported highly significant differences in nutritional aspects as a function of genetic makeup. They concluded that some hemp accessions exhibited superior nutritional attributes compared to some improved cultivars. Hence, the genetic makeup of a hemp plant will affect the nutritional value of the harvested seed. Perhaps this could be a partial explanation of the fact that hemp seed has not become more widely used as a food and source of health benefits across the millennia of human existence.

A second potential hypothesis for the somewhat limited historical and current utilization of hemp seeds and derivatives as food is competition with other foods.

Pihlanto et al. (2017) quantified the different nutritional values of several protein-rich, plant-based foods including faba bean (*Vicia faba* L.), lupin (*Lupinus angustifolius* L.), rapeseed press cake (*Brassica rapa* L./ *Brassica napus* subsp. Oleifera), flaxseed (*Linum usitatissimum* L.), oilseed hemp (*Cannabis sativa* L.), buckwheat (*Fagopyrum esculentum* Moench), and quinoa (*Chenopodium quinoa* Willd.). The authors concluded that all of the sources assayed had positive nutritional attributes and the potential to serve as useful sources of human nutrition. Interestingly, hemp seed was not superior to any of the other sources in any nutritional attribute assayed to include crude protein, amino acid profile, sugars, minerals, and dietary fibers. Hemp seed was not mentioned within their conclusions as exhibiting any superior attributes. Based on this and other studies, we can conclude that hemp has always and continues to compete with many sources of plant-derived proteins and other nutrients. In other words, relatively speaking, hemp seed is not necessarily spectacular in any regard as a source of human or animal nutrition or other health benefits. There are multiple other sources of the same nutritional components provided by hemp seeds and derivatives, and in some cases, superior sources. Hence, perhaps the peoples of Asia knew this very well and have adapted their agricultural efforts and diets accordingly. Consider the report from Bonjean et al. (2016) that in 2015, China was the second largest producer of rapeseed in the world with 18.75 million acres yielding 14.1 million tons; second only to Canada. Recall that China was reported to produce 16,424 tons of hemp seed in 2017.

We must also consider that despite the positive evidence supporting the inclusion of hemp seed or derivatives in our diet, no food source is absolutely perfect. That is certainly also the case with hemp-derived foods. Russo and Reggiani (2015) reported statistically significant concentrations of several antinutritional constituents in hemp seed meal including phytic acid (negative impact on mineral absorption), condensed tannins (multiple antinutrional effects) cyanogenic glycosides (generally minor effects at minimal doses), trypsin inhibitors (reduce protein digestion) and saponins (can negatively affect nutrient absorption). Pojicì et al. (2014) reported the same compounds, but also that fractionation of hemp seed components could isolate the nutritional and antinutritional compounds, thus providing for separate processing. This would, of course, require increased input relative to not fractionating, thus increasing the cost of the resulting foods. Of the compounds reported above, Russo and Reggiani noted that phytic acid was by far the most serious antinutrional compound present in hemp seed meal. Phytic acid was also reported as a negative attribute of hemp seed food products by Galasso et al. (2016), who suggested that reducing phytate through breeding would be a necessary genetic improvement for hemp grain to become more desirable as a food and feed. Again, perhaps ancient peoples derived this knowledge through trial and error, and adjusted their agricultural efforts and resulting diets accordingly.

Related to the fact that hemp seeds contain unfavorable nutritionally unfavorable chemicals, several workers have determined that hemp-derived proteins are certainly not the most useful or digestible. This is true considering both animal and plant protein sources. For example, House et al. (2010) reported several sources of protein exhibit protein digestibility-corrected amino acid scores (PDCAAS) superior to hemp, including casein, egg white, beef, soy protein isolate, chickpeas, pea flour, and kidney beans. Dehulled (seed coat removed) hemp seed rated highest of the hemp-derived products evaluated in their work with a PDCAAS score of 61/100. Hemp seed meal was lowest of the hemp-based products with a score of 48. Both casein and egg white have maximum scores of 100. Pea flour and kidney beans has scores of 69 and 68, respectively. Already, there is a direct effort to produce hemp protein powder to compete with other common formulations. But, at the current levels of processing, hemp protein powder is generally less concentrated than other plant-based sources, like soy and pea powders. Again, is this fact another potential reason why hemp seeds have not been utilized more widely in human agriculture and diets?

One very popular attribute of hemp seeds and derivatives that is often provided in common rhetoric is an approximate ratio of omega 6 to omega 3 fatty acids of 3 to 1. Today, it is generally accepted that average western diets contain far higher ratios; perhaps as high as 12 or 15 to 1. Further, it is hypothesized that humans evolved with diets providing ratios closer to 1 to 1, or in some cases, reversed ratios like 1 to 4. Hence, our elevated omega 6 to omega 3 ratios derived from modern western diets are thought to potentially contribute to many negative health outcomes (Simopoulos, 2016, 2002; Lands, 2014; Akerele and Cheema, 2016; among many others). Rodriguez-Levya and Pierce (2010) reported that the simple inclusion of hemp seeds in the diet can beneficially influence heart disease. Conversely, it is not perfectly clear within the modern medical literature that a reduced or reversed ratio of omega 6 to omega 3 is universally beneficial. Erkkilä et al. (2008) reported a lack of evidence that the high ratio generally found in western diets directly contributes to cardiovascular disease. This premise is supported by many in the literature (e.g., Chinello et al., 2017; Medenwald et al., 2019). Harris et al. (2006) also reported that omega 3 fatty acids alone were better predictors of coronary artery disease than the ratio of omega 6 to omega 3, and further that omega 6 fatty acids alone were not at all useful as a predictor. Additional research will ultimately answer these questions, thus providing a clearer picture of where hemp seeds and derivatives may best contribute to healthy diets.

We must not overlook the potential for hemp seeds to contribute to modern diets as additives rather than stand-alone products, as has been mostly discussed above. If we consider our current large commodity seed crops (e.g., corn and soybean), their uses as additives are far more common than uses as stand-alone products. If hemp seeds become a true commodity in U.S. agriculture, it is nearly certain that their use as food additives may become the most common form of utilization. After all, how often do we purchase and consume toasted soybeans or corn seeds? Or raw, de-hulled versions of the same? The answer is essentially never. Tofu would be an example of a popular, "stand-alone" product from soybean. Tortillas could generally be considered a stand-alone product derived from field corn.

Work has begun evaluating hemp derivatives as food additives. Chen et al. (2012) reported that an extract from hemp hulls could contribute significant antioxidant activity as an additive. Malomo and Aluko (2015) reported that both water-soluble albumin and salt-soluble globulin found in hemp seeds could be very valuable as food additives. Aiello et al. (2016) provided an excellent proteomic analysis of hemp seeds. They reported high potential for utilizing hemp seed derivatives in food products. Their work strongly contributes to understanding the molecular basis for the positive nutritional attributes possessed by hemp seed. Almost certainly, additional food science research will elucidate other, as of yet unknown uses of hemp seeds and derivatives as food additives.

Besides its less-than-perfect nutritional value, there may be other concerns with hemp-based food products. For example, due to variance in processing techniques and/or adherence to appropriate or even required processing protocols with raw seeds, it is possible for other negative events to occur. Chinello et al. (2017) reported that a two-year-old child exhibited symptoms consistent with THC poisoning after ingesting hemp seed oil (this was hemp seed oil, not the cannabinoid oils derived from female floral tissues) at the recommendation of the attending pediatrician. The intent was to strengthen the immune system of the child via hemp seed oil therapy. Ultimately, assay by GC–MS indicated that the hemp seed oil contained a 0.06% concentration of THC. Albeit a very small concentration, it was adequate to cause neurological symptoms consistent with cannabinoid poisoning following ingestion of two tablespoons of the commercially-acquired hemp seed oil daily for three weeks. Terminating the seed oil therapy rapidly resulted in alleviation of symptoms.

Discussions of hemp grain utilization as food in this work has concentrated solely on potentials as a human food. There have been investigations using hemp seed and derivatives as food for animals; mostly livestock production (e.g., Yalcin et al., 2017; Jing et al., 2017; Gibb et al., 2005). In general, these and other workers have reported positive effects of feeding hemp seeds to animals. Many of the positive effects reported are through improvements of the nutritional characteristics of the food derived from the animals (meat and eggs), and not through improved animal health or performance. Preliminary research at the University of Kentucky has indicated that mature hemp flowers containing seeds will ensile very well, and that the nutritional and digestibility values of the resulting silage is equal to or exceeds those of high-quality alfalfa hay. There are potential issues using hemp products as animal feeds that are similar to those identified by Chinello et al. (2017) above. For example, Escrivá et al. (2017) reported measureable levels of THC in cow's milk as well as in baby formula samples in Europe However, the European Food Safety Authority released a scientific opinion (Benford et al., 2015) concluding that THC in milk from feeding hemp seeds to cows was unlikely to pose a significant health risk. Again, more research will be necessary to fully define the potential for hemp grain for inclusion as a source of nutrition in animal agriculture.

There are also efforts underway today to fund research in support of hemp seed products as pet foods and treats and as per U.S. federal guidelines. Inclusion of hemp seed products in pet foods could be a highly significant market segment for farmers and processors.

Utilization in Personal Care Products

It is difficult to find unbiased data describing the use of hemp-derived ingredients in the personal products industry. Most of the available data describing the personal care products industry has been derived and provided for profit, which clearly imparts an inherent bias. There are few refereed, replicated studies of hemp oils as ingredients or potential ingredients in personal care products. Generally speaking, hemp seed oils have some positive attributes to contribute to emulsions used in personal care products. However, most of the work has not been fully validated by replicated research.

The value of the personal care industry is significant. The industry is clearly worth hundreds of billions of U.S. dollars when considering all of the market segments participating, which are generally defined as consisting of skin care, hair care, makeup or color, fragrance, and personal hygiene products. Lopaciuk and Loboda (2013) reported average annual growth in the industry as 4.5% from 1993 to 2013. They further postulated that most of the growth was due to increased markets in what are called the BRIC countries (Brazil, Russia, India, and China). These countries represent a majority of the planet's human population, and are experiencing increases in what would be defined as middle-class consumers. Lopaciuk and Loboda (2013) reported that

BRIC countries were responsible for 81% of industry growth in 2011.

Several workers (Lopaciuk and Loboda, 2013; González-Minero and Bravo-Díaz, 2018; Duran et al., 2014) have reported that interest in products containing plant-based ingredients is increasing. These would include hemp-derived ingredients. This is true within the general industry and within the cosmeceutical industry as well. Recall that cosmeceuticals are products marketed as providing both cosmetic and health benefit properties. Also recall that the word cosmeceutical is not recognized under current U.S. law.

We can report here today that there are very successful U.S. companies operating in the personal care space with products containing hemp-derived ingredients. There are clear market segments interested in purchasing and using these products, as the success of these companies illustrate. These include soaps, shampoos, lotions, salves, tinctures, among others. A very quick perusal of popular online shopping sites or general internet searches will produce dozens of available personal care products containing hemp-derived ingredients. We cannot know today what the ultimate demand for these products will become, but is it already sufficient to support more than one significant corporate entity. We can say that until 2014, essentially all of the hemp-derived ingredients were imported for U.S. personal care products. Since then, it is very difficult to ascertain how much of current U.S. hemp production is utilized in personal care products. Carus (2017) reported that a tiny fraction of hemp seed production in the EU was utilized in the personal care industry; < 0.3% of total production in 2013. Current data from the EU or data on utilization of U.S.–produced hemp grain and/or hemp grain imported to the U.S. from other countries is simply not available.

It is very difficult to define the ultimate potential for utilization of hemp-derived products in the personal care products industry. As previously stated, it is an industry worth hundreds of billions of U.S. dollars globally and is growing at significant rates. We note that current and reliable data on this form of utilization is essentially not available. This is likely due to industry inclusion of hemp-derived ingredients being relatively new, to proprietary concerns among those entities that are active in the space, and to a lack of scientific research on this form of utilization. We are proposing today that the potential for utilization of hemp-derived ingredients in the personal care products industry could very well have a significant impact on the agronomy of hemp production at least on regional scales (e.g., variety selection, crop culture, harvest timing, etc.). In other words, it is entirely possible that some hemp crops may be intentionally cultured for this purpose and in manners that could differ from the normal production protocols for cannabinoids, fiber, and grain for other purposes.

Utilization in Production of Bioenergy

As concerns about global warming continue to increase, many in science are anxious to investigate new potential renewable sources of energy. This is true within nearly all scientific disciplines, including agronomy. The current refereed scientific literature contains well over 10,000 entries reporting on feedstock utilization for bioenergy from 1977 to 2019. These efforts have included investigations of many dozens of plant species over the decades, whole or in part. More recent work has included industrial hemp.

This chapter is focused on hemp grain, so we will maintain that focus in this brief discussion. Discussion of other potential contributions from hemp toward bioenergy are included elsewhere in this book.

Das et al. (2017) reported that the inclusion of hemp seeds in calculations of the energy potential of hemp provided positive results in ethanol production. But, they also reported that the potential energy derived from hemp products is probably no more than that derived from other feedstocks. Rather, they proposed that the efficiency of hemp production was slightly higher than the efficiencies of producing other feedstocks. While this may be the case today, as is described in detail in other chapters, once pests are discovered in broad-acre hemp production systems, inputs in hemp crops (e.g., herbicides, insecticides, fungicides) will increase, and likely negate any advantages in efficiencies that hemp crops may enjoy today.

Li et al. (2010) reported that diesel fuel produced from virgin hemp oil possessed several positive characteristics (low flash point, kinematic viscosity). They also reported that virgin hemp oil produced high yields and high quality biodiesel fuel. All considered, these attributes could make virgin hemp oil competitive in this industry. Other workers (Kulglarz and Grubel, 2018; Kreuger et al., 2011) have also reported successful conversion of hemp biomass to bioenergy.

As noted by Das et al. (2017), hemp's contribution to bioenergy will be defined by its efficiencies. Plant breeding will likely contribute new hemp germplasm bred for high yields and bioenergy conversion rates. General production models will be refined also increasing efficiencies. However, there are several species of plants that serve very well as feedstocks for biofuels. The improvements noted above will determine where hemp will fit in future bioenergy models.

Summary

Grain harvested from industrial hemp may be utilized in many ways, just as are grains from many other plant species. What will define the scope and scale of a hemp grain industry is how the derived products will compete in performance and efficiencies with existing technologies, which will define ultimate consumer demands. The list of potential comparisons of hemp seeds to existing technologies is too large for a general conversation, but would include products for food, personal care products, and bioenergy. For example, hemp grain-derived products will compete with other oilseeds like flax, canola and/or rapeseed, and soybeans. Oil derived from hemp seeds has different properties than other vegetable oils, but the differences are not always positive. Hemp seed oils are generally not appropriate for cooking applications because of a relatively low flash (smoke) point. Hemp seed oils are mostly appropriate for dressing applications. This definitely limits potential utilization. The ratio of omega 6 to omega 3 fatty acids in hemp seed oils are often reported to be very desirable, but there are also reports that the ratio in flax seed oils (1 to 4) is more desirable, and that the actual effects of the 3 to 1 ratio on human health is not yet fully delineated.

The ultimate potential utilization of hemp-derived ingredients in personal care products is extremely large. The current size and growth rate of the personal care industry are simply immense. Although certainly not always supported by scientific evaluation, use of hemp-derived ingredients in many diverse personal products seems to be increasing. This could be a large area for growth as hemp becomes more widely available through increase production efforts.

Current evaluations indicate that hemp has strong competition from other feedstocks in the bioenergy sector. In other words, hemp-derived energy may be as productive and efficient as other current sources of feedstocks, but not necessarily superior. This will probably mean that hemp-derived bioenergy will likely be a secondary product (e.g., utilize the straw for cellulosic fermentation from what was originally harvested as a hemp grain crop).

As utilization of hemp-derived products is a relatively new opportunity, we often find ourselves evaluating thoughts and opinions that are not science-based. This is true at some level for all hemp-derived products. Of all of the potential derivatives from hemp crops (fiber, grain, or cannabinoids), hemp grain may be the most susceptible to what we might consider an essentially unquantifiable trait; one that could result in massive demand, hence supporting large, nationwide production efforts.

When we consider hemp-derived products for personal use (food and personal care products), the decisions of consumers to utilize these products is often supported by anecdote, rhetoric, or marketing; not often science. These are intensely personal decisions as we ingest these products and are counting on them for our sustenance and health. This is very different than deciding whether to use green energy sources or buy automobiles that contain green materials. If we accept these premises, then what drives the positive decision to use hemp-derived products if basic science is not offered in support of that decision?

I offer that our society, even more so than western society in general, has an essentially undefinable and unquantifiable connection to *Cannabis* on emotional and/or social levels. Hartig and Geiger (2018) reported that 62% of Americans support legalization of marijuana. This approval rating has trended significantly upward over time as an older generation has expired and a younger generation has become the new older generation. This means that older adults today are now more accepting of, or perhaps even through their own experiences and families they are connected to, *Cannabis* in general, to include federally-illegal marijuana. We can't fully define this connection. We certainly can't quantify it beyond polls like Hartig and Geiger reported above. Yet, it apparently exists.

This acceptance and connection is clearly evident in the industrial hemp space as well. All one needs to do to understand this is to attend an industrial hemp conference or convention. It will be abundantly clear that the populations participating in these events are generally pro-cannabis in their opinions and purchasing habits. Hence, it seems clear that a significant portion (perhaps even a majority at 62%?) could support hemp-derived products mostly because they are hemp-derived, and not necessarily because there's great science in support of that decision. The reader may immediately suppose at this point that many references from the scientific literature were referenced in this chapter regarding hemp-derived nutrition from seed among other uses. This is certainly true, but please understand also that many other sources exist containing essentially the same nutritional benefits touted in these references. They are generally accessible and not too expensive. In other words, many of these nutritional attributes are simply not unique to hemp seed. Yet, there are those that will publically suppose hemp seed is a superfood; perhaps even far more nutritious than other potential sources. Today, science just doesn't fully support that opinion.

As a scientist, I am reluctant to offer wholly unscientific opinions as above and regarding our connection(s) to hemp. But, it does seem abundantly clear that a significant portion of our society is supporting hemp-derived products even without significant science backing up those decisions. They could just as easily use flax seed in their diet and derive essentially the same benefit; yet they don't. It seems quite clear through the public rhetoric surrounding many hemp-derived products that this support is mostly due to the genus to which the species belongs, and not necessarily to refereed reports in the scientific literature supporting superior performance attributes relative to other plant species.

That said, it probably doesn't so much matter in the big picture. By most accounts, hemp seeds are indeed tasty. As far as we know today they are not unhealthy in any way unless contaminated with foreign, unwanted compounds; however, that's true of any food. It is not yet clear today what hemp-derived ingredients may contribute to the personal care products industry, but the potential there is nearly mind-boggling. If only a third of products ever contained hemp-derived ingredients; that would involve several billions of U.S. dollars. And lastly, most everyone wants greener energy. Hemp can and probably will contribute to that reaching that goal, but it may or may not be utilized predominantly relative to other feedstocks.

From the perspective of agronomic science, we are working very hard to support the evolving industry through the sound, scientific definition of the most efficient and profitable production models so that no matter the motivation, consumers may continue to expand their demands for hemp-derived products. It is truly a very exciting time to be involved in hemp grain production research.

Literature Cited

Aiello, G., E. Fasoli, G. Boschin, C. Lammi, C. Zanoni, A. Citterio, and A. Arnoldi. 2016. Proteomic characterization of hempseed (Cannabis sativa L.). J. Proteomics 147:187–196. doi:10.1016/j.jprot.2016.05.033

Akerele, O.A., and S.K. Cheema. 2016. A balance of omega-3 and omega-6 polyunsaturated fatty acids is important in pregnancy. Journal of Nutrition & Intermediary Metabolism 5:23–33. doi:10.1016/j.jnim.2016.04.008

Benford, D., S. Ceccatelli, B. Cottrill, M. DiNovi, E. Dogliotti, L. Edler, P. Famer, P. Fürst, L. Hoogenboom, et al. 2015. Scientific opinion on the risks for human health related to the presence of tetrahydrocannabinol (THC) in milk and other food of animal origin. EFSA Journal. 13:4141. doi:10.2903/j.efsa.2015.4141.

Bonjean, A.P., C. Dequidt, and T. Sang. 2016. Rapeseed in China. OCL. 23: D605. doi:10.1051/ocl/2016045. doi:10.1051/ocl/2016045

Callaway, J.C. 2004. Hempseed as a nutritional resource: An overview. Euphytica 140:65–72. doi:10.1007/s10681-004-4811-6

Carus, M. 2017. The European hemp industry: Cultivation, processing and applications for fibres, shivs, seeds and flowers. EIHA 2017-03. http://eiha.org/media/2017/12/17-03_European_Hemp_Industry.pdf (Accessed 13 Mar. 2019).

Chen, T., J. He, J. Zhang, X. Li, H. Zhang, J. Hao, and L. Li. 2012. The isolation and identification of two compounds with predominant radical scavenging activity in hempseed (seed of Cannabis sativa L.). Food Chem. 134:1030–1037. doi:10.1016/j.foodchem.2012.03.009

Chikezie, P.C., and O.A. Ojiako. 2015. Herbal medicine: Yesterday, today and tomorrow. Altern. Integr. Med. 4:3. doi:10.4172/2327-5162.1000195

Chinello, M., S. Scommegna, A. Shardlow, F. Mazzoli, N. De Giovanni, N. Fucci, P. Borgiani, C. Ciccacci, A. Locasciulli, and M. Calvani. 2017. Cannabinoid poisoning by hemp seed oil in a child.

Pediatr. Emerg. Care 33:344–345. doi:10.1097/PEC.0000000000000780

Das, L., E. Liu, A. Saeed, D.W. Williams, H. Hu, C. Li, A.E. Ray, and J. Shi. 2017. Industrial hemp as a potential bioenergy crop in comparison with kenaf, switchgrass and biomass sorghum. Bioresour. Technol. 244:641–649. doi:10.1016/j.biortech.2017.08.008

Duran, I., A. Bikfalvi, and J. Llach. 2014. New facets of quality. A multiple case study of green cosmetic manufacturers. European Accounting and Management Review 1:44–61. doi:10.26595/eamr.2014.1.1.3

Erkkilä, A., V.D.F. de Mello, U. Risérus, and D.E. Laaksonen. 2008. Dietary fatty acids and cardiovascular disease: An epidemiological approach. Prog. Lipid Res. 47:172–187. doi:10.1016/j.plipres.2008.01.004

Escrivá, Ú., M.J. Andrés-Costa, V. Andreu, and Y. Picó. 2017. Analysis of cannabinoids by liquid chromatography–mass spectrometry in milk, liver and hemp seed to ensure food safety. Food Chem. 228:177–185. doi:10.1016/j.foodchem.2017.01.128

FAO. 2019. FAOSTAT Crops. Food and Agriculture Organization, Rome, Italy. http://www.fao.org/faostat/en/#data/QC (Accessed 13 Mar. 2019). [2019 is year accessed].

Galasso, I., R. Russo, S. Mapelli, E. Ponzoni, I.M. Brambilla, G. Battelli, and R. Reggiani. 2016. Variability in seed traits in a collection of Cannabis sativa L. genotypes. Front. Plant Sci. 7: 688. doi:10.3389/fpls.2016.00688.

Gale, R., and G. Null. 2018. China's botanical medicine. The reality of natural medicine. China's traditional herbology. Global Research, Montreal, Canada. https://www.globalresearch.ca/chinas-botanical-medicine-the-reality-of-natural-medicine/5647242 (Accessed 13 Mar. 2019).

Gibb, D.J., M.A. Shah, P.S. Mir, and T.A. McAllister. 2005. Effect of full-fat hemp seed on performance and tissue fatty acids of feedlot cattle. Can. J. Anim. Sci. 85:223–230. doi:10.4141/A04-078

Harris, W.S., B. Assaad, and W. Carlos Poston. 2006. Tissue Omega-6/Omega-3 fatty acid ratio and risk for coronary artery disease. Am. J. of Cardiology (www.AJConline.org) Vol. 98: (4A). doi:10.1016/j.amjcard.2005.12.023.

Health Canada, 2019. Statistics, reports, and facts sheet on hemp. Government of Canada, Ottawa, ON. https://www.canada.ca/en/health-canada/services/drugs-medication/cannabis/producing-selling-hemp/about-hemp-canada-hemp-industry/statistics-reports-fact-sheets-hemp.html (Accessed 3 Mar. 2019). [2019 is the year accessed].

House, J.D., J. Neufeld, and G. Leson. 2010. Evaluating the quality of protein from hemp seed (Cannabis sativa L.) products through the use of the protein digestibility-corrected amino acid score method. J. Agric. Food Chem. 58: 11801–11807. doi:10.1021/jf102636b.

Jing, M., S. Zhao, and J.D. House. 2017. Performance and tissue fatty acid profile of broiler chickens and laying hens fed hemp oil and HempOmega. Poult. Sci. 96:1809–1819. doi:10.3382/ps/pew476

Johnson, R. 2018. Hemp as an agricultural commodity. 7-5700. RL32725. Congressional Research Service, Washington, D.C. www.crs.gov (Accessed 5 Mar. 2019).

González-Minero, F.J., and L. Bravo-Díaz. 2018. The use of plants in skin-care products, cosmetics and fragrances: Past and present. Cosmetics 5:50. doi:10.3390/cosmetics5030050

Kreuger, E., T. Prade, F. Escobar, S.-E. Svensson, J.-E. Englund, and L. Björnsson. 2011. Anaerobic digestion of industrial hemp: Effect of harvest time on methane yield per hectare. doi:10.1016/j.biombioe.2010.11.005.

Kulglarz, M., and K. Grübel. 2018. Integrated production of biofuels and succinic acid from biomass after thermochemical pretreatments. Ecol. Chem. Eng. S 25:521–536. doi:10.1515/eces-2018-0034

Lands, B. 2014. Historical perspectives on the impact of n-3 and n-6 nutrients on health. Prog. Lipid Res. 55:17–29. doi:10.1016/j.plipres.2014.04.002

Li, S.-Y., J.D. Stuart, Y. Li, and R.S. Parnas. 2010. The feasibility of converting Cannabis sativa L. oil into biodiesel. Bioresour. Technol. 101:8457–8460. doi:10.1016/j.biortech.2010.05.064

Łopaciuk, A., and M. Łoboda. 2013. Global beauty industry trends in the 21st century. Proc. of the Active Citizenship by Knowledge Management and Innovation, Management, Knowledge and Learning International Conference. 19-21 June Zadar, Croatia. 1079–1087.

Malomo, S.A. and R.E. Aluko. 2015. A comparative study of the structural and functional properties of isolated hemp seed (Cannabis sativa L.) albumin and globulin fractions. Food Hydrocolloids. 43:743–752. doi:10.1016/j.foodhyd.2014.08.001.

Medenwald, D., A. Kluttig, M.E. Lacruz, and J. Schumann. 2019. Serum dietary fatty acids and coronary heart disease risk-A nested case-control-study within the CARLA cohort. Nutr. Metab. Cardiovasc. Dis. 29:152–158. doi:10.1016/j.numecd.2018.10.006

Pihlanto, A., P. Mattila, S. Mäkinen, and A.-M. Pajari. 2017. Bioactivities of alternative protein sources and their potential health benefits. Food Funct. 8:3443–3458. doi:10.1039/C7FO00302A

Pojicì, M., A. Mišan, M. Saka , T. Dap evi Hadnadev, B. Šaricì, I. Milovanovi , and M. Hadnaðev. 2014. Characterization of byproducts originating from hemp oil processing. J. Agric. Food Chem. 62: 12436-12442. https://pubs.acs.org/doi/10.1021/jf5044426.

Rodriguez-Leyva, D., and G.N. Pierce. 2010. The cardiac and haemostatic effects of dietary hempseed. Nutrition & Metabolism 7:32 http://www.nutritionandmetabolism.com/content/7/1/32. doi:10.1186/1743-7075-7-32

Russo, R. and R. Reggiani. 2015. Evaluation of protein concentration, amino acid profile and antinutritional compounds in hempseed meal from dioecious and monoecious varieties. Am. J. Plant Sci. 6:14–22. doi:10.4236/ajps.2015.61003

Small, E. 2015. Evolution and classification of Cannabis sativa (marijuana, hemp) in relation to human utilization. Bot. Rev. 81:189–294.

Simopoulos, A.P. 2002. The importance of the ratio of omega-6/omega-3 essential fatty acids. Biomed. Pharmacother. 56:365–379. doi:10.1016/S0753-3322(02)00253-6

Simopoulos, A.P. 2016. An increase in the Omega-6/Omega-3 fatty acid ratio increases the risk for obesity. Nutrients 8:128. doi:10.3390/nu8030128

Yalcin, H., Y. Konca, and F. Durmuscelebi. 2017. Effect of dietary supplementation of hemp seed (Cannabis sativa L.) on meat quality and egg fatty acid composition of Japanese quail (Coturnix coturnix japonica). J. Anim. Physiol. Anim. Nutr. 102:131–141. doi:10.1111/jpn.12670

Chapter 3: Hemp Fibers

Trey Riddle,* Jared Nelson, and Patrick Flaherty

Introduction

It is rare that a new crop can be successfully introduced into rotations dominated by traditional commodity crops. Specialty crop contracts enable farmers' opportunities to grow incrementally more value crops than corn, soybeans, and wheat. Effectively adding biodiversity to current agricultural production models is difficult due to an immediate lack of profitability. Integrating new crops on a small, research scale is imperative to show the beneficial impact on both the soil and farmers' lives through quantitative analyses. Providing farmers more opportunities to grow a diversified crop rotation portfolio is proven to improve soil health and the surrounding environment while spreading the fiscal risk of crop production (Cothren, 2014; Government of Alberta, 2004; Penn State College of Agricultural Sciences Cooperative Extension, 1996).

Field-scale production of industrial hemp in order to meet the high value, emerging markets for biomaterial or bioindustrial applications is underway to the benefit of entire value chain. These new industrial crops will potentially cover thousands of acres of farmland as well as add substantial numbers of new jobs. Data generated by the USDA affirms this position by showing that for every one job created in biobased companies, an additional 1.64 more jobs are created outside of the company (Golden et al., 2015).

With natural fibers, end-user products will play the biggest role towards benefiting our society. Consumer demands are leaning towards more natural-based products as seen with large companies (i.e. Ford and Walmart), shifting towards greener initiatives. Social benefits correlate directly with environmental initiatives as a better physical environment can lead to better social environment. Much like the industrial revolution benefitting society and spurring a more advanced world, the revolution of creating an environmentally conscious world will contribute directly towards healthier global growth.

Even with the rapid growth to date of the emerging industry and new agricultural regions under hemp production, we do not yet have the historical data necessary to identify the optimal cultivars, growing conditions and inputs, types of processing, processing parameters, and materials specifications for the market to achieve optimal success. For successful development of a hemp fiber industry, the feedstock supplied by the farmer needs to be used almost entirely, with

T. Riddle and P. Flaherty, Sunstrand, LLC, Louisville, KY 40206; J. Nelson, State University of New York, New Paltz, NY 12561. *Corresponding author (triddle@sunstrands.com)

doi:10.2134/industrialhemp.c3
Industrial Hemp as a Modern Commodity Crop. D.W. Williams, editor.

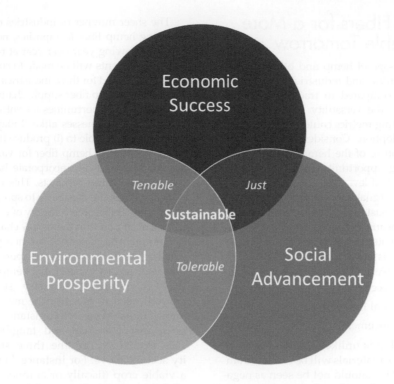

Fig. 1. Sustainability is realized when all three sectors are simultaneously achieved.

little to no material waste. This maximizes income for farmers and generates an economic boost from products downstream in the supply-chain. Without developing a complete understating of these materials and their compatibility while also assessing values, these opportunities will be lost, or perhaps never even defined. As such, a comprehensive understanding of hemp and other bio-industrial materials, and their potential opportunities, is needed to increase competitiveness and material use relative to incumbent traditional materials.

To maximize their income, farmers need to supply feedstocks generating significant economic return from products downstream in the supply-chain. Technical industrial materials generally rely on the bast stalk which consists of both the outer bast fiber and the inner woody core commonly known as hurd. Processors have begun to implement methods of extraction to produce a high quality technical fiber for industrial applications (e.g. yarns, composites, textiles, etc.), and while also expanding the technologies capable of utilizing hurd material. The carbon rich inner woody core is now seen as a viable substrate for multiple advanced technical applications such as composites and micro and nano cellulose applications.

It is estimated that natural fibers require approximately 70% less energy to process and have 75% less CO_2 emissions compared to fiberglass (Hockstad and Weitz, 2015). However, for industry to fully utilize natural fibers, fiber and *in situ* (e.g. application or part) properties, such as strength, stiffness, flexibility, wear, and abrasion, need to be better understood as functions of both agronomic and processing variables. Physical fiber properties are difficult to ascertain due to the variation in the geometry and structure of these fibers meaning strength can be difficult to measure. The natural variability of hemp fiber and hurd can be characterized by traditional metrics and the uncertainty quantified. However, hemp materials are most commonly used in conjunction with other materials (e.g. composites) where the blending of materials generally results in a consistent application response. Moreover, the real success of hemp fibers in industrial and technical applications is most commonly a function of how it compares to other plant-based and synthetic materials on cost and performance bases.

Natural Fibers for a More Sustainable Tomorrow

At present, usage of hemp and other natural fibers in technical and industrial applications is negligible compared to traditional materials. However, the versatility of hemp fibers and their costing metrics could result in major industrial adoption. Considering the underdeveloped nature of the hemp fiber industry, an interesting opportunity exists to create a supply chain that leverages current consumer affinity (and arguably societal necessity) for sustainability. Sustainability can have a variety of different meanings based on context. In the context of a material supply we can delineate sustainability as the nexus of three sectors, as adapted from (Asby, 2009) and shown in Fig. 1:

- Economic Success
- Environmental Prosperity
- Social Advancement

In general, it is unlikely that any wholly new or unique materials will be created from hemp fiber. This should not be seen as negative; quite the contrary, a massive industrial base has developed over the last 100 years around the utilization of synthetic materials. The replacement of these materials with hemp fiber will constitute a double benefit between plant growth characteristics (e.g. CO_2 sequestration) and the supplanting of energy intensive materials that utilize pollution generating manufacturing processes.

The sheer number of industries capable of adopting hemp fiber (composites, nonwovens, etc.) are growing year over year at rapid rates. New investments will be made to support the materials needed for these industries. However, the lack of a hemp fiber supply chain provides demonstrable opportunities for entrepreneurs and existing businesses alike. Today, the economics are favorable to (i) produce hemp fiber crops, (ii) process hemp fiber for value-added applications, and (iii) incorporate hemp fiber into manufactured products. This new financial capital has the capability to spread wealth and sustainability into a variety of demographics and at all points in the supply chain.

Though the potential for hemp fiber is exciting, it must also be tempered with a few realities. In general, for hemp to reach the full potential, industry stake holders will need to identify the most relevant applications and circumstances where hemp provides real and tangible values. This extends to all the three sustainability sectors above. For instance, hemp is not a viable crop (fiscally or agronomically) in all environments, therefore the most appropriate regions will need to be identified for cultivation and processing. Moreover, it is important to realize that there must be a balance between current industrial and social norms and the concept of disruption. Take for example the need to improve the sustainability of transportation (Fig. 2). If we were to promulgate the ultimate sustainable transportation solution, we would likely

Fig. 2. Finality of a thought experiment where the end goal is to be as sustainable as possible (Permission: Sunstrand).

Table 1. Energy estimations for fiberglass (Dai et al., 2015) and Sunstrand biomaterials production.†

	Fiberglass	Natural Fibers	With Hurd Production
Energy Usage (MMBtu ton^{-1})	12.93	4.02	0.670
CO_2 Emission from Electrical Usage (lb ton^{-1})*	1512	470	78.4
CO_2 Emissions during Production - Melting Refining (lb ton^{-1})	348	0	0
Total CO_2 Production Emissions (lb ton^{-1})	1861	470	78.4
Water (gal ton^{-1})	960	0	0

† Based upon a 117.0 lb CO_2 MMBtu^{-1} released during energy production from natural gas

end up either on horses or densely packing people onto trains. It is easy to see how these types of changes would result in extreme preservation of the environment, but at the same time would significantly (and most would say negatively) affect our way of life.

Examples of Impacts when Replacing Select Traditional Materials

Bast plants sequester carbon dioxide (CO_2) out of the atmosphere which accounts for 70 to 80% of greenhouse gases (GHG) (U.S. Environmental Protection Agency, 2017). Emission studies conducted by the U.S. government have found that the transportation sector is the second largest emitter of GHG (Hockstad and Weitz, 2015). Natural fibers may reduce the weight of some component products relative to synthetic fibers in the transportation sector. Reduced weight could contribute to increased efficiencies in all facets of the transportation industry. The growth of an industry utilizing plant-based materials will help mitigate GHG emissions during processing and production. Moreover, when utilizing agricultural materials as feedstocks for advanced technologies, any adverse effects of production will at least in part be offset during the growing season by normal carbon sequestration supporting plant growth and development. In short, if transportation is made more efficient by natural fibers, and if processing natural fibers contribute less GHGs than synthetic fibers, and fiber crops sequester carbon annually during the growing season, it follows logically that natural fibers provide an environmental

advantage over synthetic fibers when utilized in the same applications.

Current use of resources in the production of fiberglass and processing of natural fibers can be seen in Table 1 below. This data exemplifies the use of non-renewable natural resources for the production of electrical grade glass fiber for the production of composite materials. A total of 12.93 MMBTU is required for the production of a ton of glass fiber, as well as using 960 gallons of water per ton. This water is used for cooling the glass and is evaporated during production creating a loss. Also, not accounted for is the transportation of raw materials to the production plant, releasing a significant amount of CO_2.

Utilizing the U.S. Energy Information Administration (2015) carbon dioxide emission data for natural gas, an estimation regarding CO_2 emissions from electrical usage and production and refining processes can be made [9]. Production of composite grade natural fibers requires only around 4.02 MMBtu ton^{-1} of energy. These values are based on fiber output for comparison with glass fiber, but also produce hurd material for bio-based applications. When the mass of hurd material is included, energy consumption drops down to around 0.670 MMBtu ton^{-1}. Concurrently, CO_2 emission values increase by of 30% due to the extra machines required for processing but result in lower emissions per ton as output of product is increased five-fold. Depending on crop yields, approximately 8000 acres of farmland is required to support one of major industrial fiber processing facility with a throughput of approximately 3 tons hr^{-1}. The values provided in Table 1 do not reflect the additional 1.796 tons of CO_2 sequestered per acre during a growing season, increasing the net amount of CO_2 offset during production (Vosper, 2011).

Industry and Advancements

Ubiquitous utilization of hemp fiber will require focused and advanced developments in key areas. As the resources are deployed, industry growth will expand from easy, readily available applications to exotic applications. In doing so, all facets of materials development will increase including the complexity of the material, value of the material, and timeline to adoption. To begin the conversation around hemp fiber applications we can start with a high-level grouping into the general areas:

- Fiber (elongated, high aspect ratio)
- Particulates (granular, chips, fines)
- Other (converted raw materials)

Within each of these categories are a variety of potential applications. The following chart (Fig. 3) begins to illustrate the wide-reaching potential of the hemp stalk. There are several overlapping material architectures and applications. In time, industry should endeavor to flush a map of these applications overlaid with value and application compatibility to ensure resources are being deployed in places that make the most sense.

It is also interesting to note that much of the rest of the world has begun (or never stopped) using natural fibers, whereas the United States is behind (J.E.C. Composites Publications, 2014). One could point to this fact as disconcerting, but with further introspection, the conclusion can be reached that there is a massive opportunity for the natural fiber industry to grow within the world's largest economy. One market which is particularly exciting is composites. While broad in nature, a critical nexus can be described in which there is (i) a fast-growing industry, (ii) hemp fibers have relevant material properties, (iii) price points are cost effective, and (iv) there is significant, drop-in compatibility with existing manufacturing infrastructure.

Summary of Introduction

High value, emerging markets for biomaterial or bioindustrial applications are gaining traction to the benefit of entire value chain. The materials from both the bast fiber and hurd portions of hemp stalk offer many technical application opportunities, though a more comprehensive understanding is needed to ensure the adoption, competitiveness, and ultimate usage in new market segments. However, many existing opportunities can use hemp at cost and performance parity or advantage while leveraging current consumer affinity for sustainability, particularly when considering replacement of existing traditional materials. To achieve billion-dollar industry utilization, material development is necessary in value and application compatibility to ensure resources are being deployed effectively. A discussion

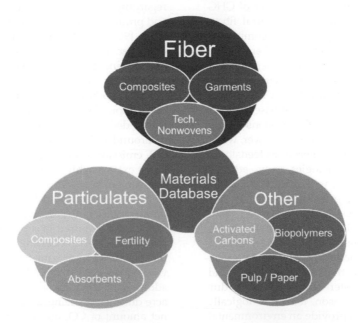

Fig. 3. Map of material systems and select applications (Permission: BMI/Sunstrand).

Fig. 4. Image of hemp stalk showing the bast fiber outer and woody core inner components (Permission: Sunstrand).

that hemp is very similar to a wide variety of plants. More specifically, bast fibers are collected from the fibrovascular bundle portion of the stalk commonly referred to as the bast. These fibers are located between the protective epidermis and the inner woody core. The long fiber makes up the majority (70-90%) of the bast and generally has high cellulose and low lignin content resulting in fine, flexible fiber, making it valuable once extracted. The remainder of the bast consists of shorter, more lignified fibers, sometimes referred to as bark.

Lignin is a class of complex, cross-linked phenolic polymers that play a big role in the rigidity of plant cell walls, especially in wood and bark. They are relatively hydrophobic and aromatic in nature. Their structures are very complex and have not been fully identified to this day (Vuorinen, 2019; 13). Cellulose is the most abundant organic polymer in the world and is used in a variety of applications such as paper products, cellophane, rayon, dietary fiber, and biofuels. It is a linear polymeric chain that is insoluble in water and in most organic solvents but is biodegradable. The intra- and intermolecular hydrogen bonds through the free hydroxyl groups (OH) present in each of the repeating units of the cellulose chain cause the molecules to orient in an orderly manner predominantly parallel to each other. This hydrogen bonding allows the chains to form microfibrils with a high tensile strength. These microfibrils are meshed into a polysaccharide matrix within the cell wall, creating a crystalline structure. Many properties of cellulose depend on its degree of polymerization (the number of glucose units that make up one polymer molecule). Cotton and other plant fibers have chain lengths ranging from 800 to 10000 units.

of the state-of-the-art growth and processing surrounding hemp straw with a focus on utilization follows in this chapter.

Straw and Stalk Morphology

Hemp straw consists of the stalk component of the plant (Fig. 4). Hemp is lignocellulosic in that the majority of its make-up is lignin and cellulose. In general, this is a very common form of plant matter. Table 2 provides a comparison of some select plants and their major constituent properties. It can be seen

Table 2. The chemical composition of hemp and other natural fiber producing plant species (J.E.C. Composites Publications, 2014).

Substances (% of dry matter)	Hemp	Flax	Jute	Ramie	Sisal	Abaca	Coir	Cotton
Cellulose	70	70	65	72	66	60	40	90
Hemicellulose	16	17	15	14	12	21	0.2	4
Lignin	6	2.5	10	0.7	10	10	43	0.7
Pectin	1	2	1	2	2	0.8	3	4
Fat/wax	0.7	1.5	0.5	0.3	1	1.4		0.6
Ash	1.5	1.5	0.4	0.3	0.3			1.4
Water solubles	1	6	1	6	3.5	1.4	4.5	0.7

Hemp fibers also consist of long chain, high molecular weight cellulose molecules bound with a matrix of lignin reinforcement. Within the cell wall, this cellulose is also embedded with polysaccharides, including hemicellulose, which combine to form microfibrils that build the structure of the fiber. While each fiber is the individual cell, they are bundled together with various polymeric substances. It is important to note that it is the fiber bundles that achieve the long length (1-5 m) found in hemp, but the individual fibers (cells) are much shorter (1-5 cm). The individual cells are cylindrical with a thick wall and are polygonal in cross-section with many surface irregularities. As an example, hemp fiber cells are thicker, and the central lumen is wider than that of flax and hemp fibers are stiffer than flax as they are more lignified (J.E.C. Composites Publications, 2014).

The organization and orientation of the microfibrils are the controlling factor in the performance characteristics of the plant fiber (Fig. 5). While the primary outer wall of the fiber consists of a random arrangement of the microfibrils, the secondary wall consists of three layers where the microfibrils are arranged. Specifically, in the first and third layer they are arranged in a spiral orientation; one right-handed and the other left-handed. The middle layer which forms the bulk of the secondary wall, has microfibrils aligned parallel to one another in a single steeply inclined helix. It is the angle of this helix with respect to the length of the fiber that drives the performance characteristics. As an example, the angle for hemp is much less than for cotton meaning that while the strength is increased, cotton has a much higher elongation at break.

Historically, hemp has been grown primarily for the long bast fibers. The inner woody core, commonly called hurd or shive, was a waste product. Considering the value of bast fibers and the competition from other fiber systems in the present day, an income stream based solely on bast fibers is not valid. Since each stalk is 70 to 80% hurd, a significant number of uses have necessarily been identified for this material in support of a profitable business model. The hurd is 20 to 30% lignin and has a relatively very short fiber length. Processing of the hemp stalk may follow a variety of approaches for each of these components to be made into specific value-added materials.

Relevant Agronomic Factors

A thorough discussion on hemp fiber agronomy is provided in Chapter 4. In general, growing hemp for fiber applications fits very well into traditional crop rotations such as corn and soybeans. Moreover, the planting of hemp for fiber utilizes common production processes and equipment, for example, cereal drills. Harvesting of the hemp stalk or straw can be challenging considering plant heights of up to 15 ft (5 m). However, the most critical agronomic outcome is quality. There are many factors related to production which affect fiber quality, and ultimately value, in the context of fiber utilization and processing. While various processes will have differing options, the factors affecting quality, in order of criticality during production and harvesting are as follows:

1. Degree of retting at baling

2. Maturity of plant at harvest

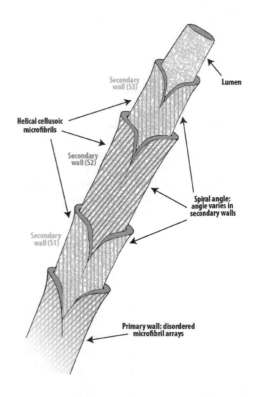

Fig. 5. Theoretical representation of the structure and arrangement of plant fiber. (Permission: BMI/Sunstrand).

3. Stem diameters and length, harvesting type and format, and material storage.

4. Variety selection

While these are the critical factors affecting fiber quality, it is easier to logically discuss them in the chronological order of production and harvesting. As such, discussion of the important points surrounding *Stem Size and Variety Selection, Maturity, Retting*, and *Harvesting Technologies* with their impact on fiber quality follows.

Stem Size and Variety Selection

Size or diameter of the hemp straw is important for two basic reasons; fiber yield and compatibility with decortication processes. In general, the larger the diameter of the stalk, the lower the bast content proportionally compared with the hurd. Conversely, the smaller the diameter, the higher the bast fiber content. As bast fiber is more valuable per pound than hurd, processors and growers will tend to optimize plant density and varietal selection to maximize bast fiber production. Plant densities (populations) will have a direct impact on stem diameters. The total biomass may be provided by large individuals resulting in fewer individuals per unit area, or conversely, the biomass may be provided by smaller individuals, allowing for many more individuals to co-exist in the same unit area. Research at the University of Kentucky has investigated the effects of hemp and kenaf seeding rates (plants per unit area) on stem diameter and the uniformity of stem diameters. The results quantified the fact that there are no significant differences in total dry matter yields (total biomass production) among seeding rates of 20, 40, or 60 pounds of seed per acre (Williams et al., 2017a,b, unpublished). However, there are significant increases in both the stem diameters and in the variation of stem diameters at 20 pounds per acre compared to 40 or 60 pounds per acre. There were no significant differences comparing 40 to 60 pounds per acre seeding rates. Proper varietal selection is necessary to address adaptation to local latitude as well as soil and climate conditions to ensure proper plant health and yields. There is minor variation in the quality of bast fibers as a function of genetic selection (variety); however, this is generally of second order impact to quality and value compared with the other factors in this section.

Maturity

Hemp plants, like all other annual plants, follow the typical life-cycle of germination, vegetative growth, and reproduction (flowering and seed setting). When harvesting hemp for grain or CBD, the common practices are to let the plant mature to flowering and seed set. Most purpose-grown fiber crops are harvested at end of vegetative growth and at the beginning of flowering. The reasons for this are two-fold; (i) yield of straw is effectively at its maximum at the end of the vegetative period and (ii) fiber quality degrades during the plant reproductive phase. The former is obvious in that once entering reproduction, the plant focuses more energy on the flower or seed head and minimal stalk growth happens (i.e., growth is determinant). However, the latter is little less obvious. Whether dioecious or monoecious, the hemp plant produces a very large female flower and seed head. When mature and/or wet, this flower is very heavy. In order to support the seed head, the stalk goes through some morphological changes. The outer bast fibers begin to bind together through the cementing of additional pectins and lignins. While this change results in a plant that is better engineered to support the seed head to prevent lodging, it also has the unfortunate side effect of making the fiber less fine, more lignified, and more challenging to separate induvial fiber bundles, also referred to as opening individual fibers.

Fiber-specific hemp varieties are typically bred to have high mechanical fiber properties, long fiber lengths, as well as high processing yield. These varieties are determined regionally due to local conditions (e.g. latitudinal and environmental). Currently, there is also great potential interest in dual-use crops for fiber production. The dual-use crops are grown primarily for the flower or grain portion of the plant and thus are not optimized for fiber production. The processing of these feedstocks differs depending on the portion of the plant being utilized and for what purpose. The format of the residual biomass thus differs as well. Despite this, there are some applications being investigated where a lower quality raw material could not only be acceptable from a processing standpoint but provide a financial advantage as well.

The general process of decortication separates the bast fiber from the hurd. Once the bast fiber is liberated, it still must be of a format which is compatible with the upstream processing and applications. Considering that the vast majority of fiber conversion technologies were developed around synthetics and cotton, they have been optimized over decades for fine fibers of a relatively consistent length. In addition, the new secondary bonds formed during reproductive growth are not as strong as the primary bonds developed during vegetative growth. Therefore, the resulting fiber bundles, if not perfectly incapsulated or engaged along the majority of their surface area, have a reduced strength. While further processing can break these bundles into finer fibers, the result is a more fractured or fibrillated fiber with shorter length with increased cost relative to fibers derived from a purpose-grown crop.

Retting

The mechanical properties of a natural fiber are highly dependent on many naturally occurring factors. These include the climate where it is grown, how and when it is harvested, and the degree to which the crop is retted. Retting, sometimes referred to as degumming, is a microbial process during which mostly bacteria digest the substances that hold fibers together, making it easier to separate them into single fibers mechanically separated from the hurd (Fuqua et al., 2012). The degree of retting will have major impact on the processability of the straw. Straw that has minimal retting will be very challenging to decorticate efficiently. Fiber that is over-retted will begin to deteriorate, losing strength and reducing yields of valuable fiber during the separation process.

The most common retting process utilized in developed countries called 'dew' or 'field' retting. In this process, the harvested stems are left on the ground for several weeks for the pectin to be digested by bacteria, enabling the ability to separate fibers during mechanical processing (Fuqua et al., 2012). Today, it appears that bacterial species involved in pectin degradation may be related

to specific hemp varieties rather than endemically present in the soil. Research to define the microbiome and microbial ecology of retting continues today. Monitoring during the three-to-five-week retting process is necessary to ensure that the material is not over- or under-retted. Given that retting condition is a function of air, humidity, and temperature, rotation of the straw is required. A significant benefit to leaving the stalks in the field during dew retting is that the process returns many of the nutrients back into the soil. Although the process is simpler than other methods, the outcome has relatively inconsistent results and can often result in reduced fiber strength. It is very difficult for even an expert to ret an entire crop to the same degree. Stem diameters are also an important aspect of uniform retting. Larger stems will rett more slowly than smaller stems. This is another example of the effects of uniform stem diameters on ultimate fiber quality.

Retting has a large contribution to the mechanical properties of the fiber. If the retting process is not done long enough, or for too long, it can damage the performance of the fiber, causing it to have lower mechanical properties. Retting also plays a large role in the color of the fiber. In fact, the degree of retting can in many cases be qualified based on the stalk color. In applications that require bleaching or dyeing (e.g. apparel), the color of the fiber affects the ease of these processes.

Water retting is an accelerated method and produces more uniform, high-quality fiber. Stacks of cut stalk are immersed in water and are monitored frequently. Water must be kept at consistent temperature and be circulated uniformly through the mass of material. This process is effective, but is also costly as is uses large volumes of clean water that must be treated before being discharged. This method has been adapted over time to include the use of chemicals to facilitate and better control the retting process. This type takes about half the time of dew retting but produces fibers of greater quality and uniformity.

Modified wet retting procedures, such as ultrasonic, steam explosion, enzymatic, and chemical retting have been developed to obtain fine, clean, consistent quality fibers (Fuqua et al., 2012). Ultrasound retting obtains fibers for non-textile applications and is more consistent than biological retting and mechanical retting. Steam explosion can produce fibers that are comparable to cotton fibers in terms of fineness and performance. Enzyme retting yields strong, fine, and consistent quality fibers. Like enzyme retting, chemical retting is a wet process that uses heat to reduce the time associated with retting. In this process, chemicals are used to degum the pectin from between the bast and hurd fibers. Both enzymatic and chemical retting are significantly more costly than field retting.

As noted above, retting has the potential to have highly significant impacts on the yields of high-quality bast fibers. Among all factors within the process that are even marginally controllable, obtaining an appropriate level of retting is either imperative to success, or a reason for crop failure. The ultimate importance of appropriate retting simply cannot be overstated. That said, teaching a farmer how to appropriately ret a hemp crop is not unlike teaching how to produce very high quality hay. We can teach farmers to mow the crop at a specific growth stage and then evaluate it at certain points in time. We cannot directly teach (or even write) what a high quality hay crop looks like, smells like, or feels like. The same is true for retting hemp crops. Appropriate retting will be quickly and very successfully learned by savvy farmers keen to succeed in hemp farming. It will be a somewhat inherent skill; perhaps as much an art as a science.

Harvesting Technologies

The selection of harvesting technology is generally dependent upon the processing technology to be used for decortication. Though there are major and notable exceptions, the majority of the current hemp straw processors require the hemp straw to be delivered in bales. Bales offer ease of integration into mainstream agronomic production models, as well as good packing density of material for transport and storage. To a lesser degree, there are processors that utilize in-field decortication and storage in a manner more similar to forage or silage.

When harvesting to package in bales, there are two main mechanical systems for hemp; whole stalk harvesting and cut stalk harvesting. Since hemp plants tend to be two to three meters taller than most other bast fiber plants, it is difficult to harvest the whole stalk without specialty equipment.

Whole stalk harvesting generally offers growers the potential to utilize existing haying equipment. For any plant over over seven feet high, it is basically necessary to utilize a sickle bar or custom chopper (forage harvester). Sickle bars are effective and common but can be relatively slow. Turning of the straw is critical to achieve uniform retting. When the stalks are relatively short (less than around five feet) a standard hay tedder can be used. When stalks are longer a power or rotary rake is most effective. Inversion rakes are ideal due to uniformity that can be achieved. Turning is typically performed two to three times through the retting process. Baling should be performed when the straw is less than 14% moisture content. Considering the long length of whole cut straw, it is most efficient to use a silage baler. These balers have knives in the front which can cut the hemp stems to lengths up to about 1.5 ft (0.4m). Cutting of the straw helps to minimize wrapping in the bale chamber; however, wrapping is still a major concern. Even within the same type of baler (i.e., round or square), some manufacturers' models will work better than others. Although specialized harvesting equipment is used in other parts of the world, fiber processors in North America work with farmers to accept bulk straw in formats that utilize existing harvesting equipment. These are typically in the form of 500 lb to 1100 lb square or round bales whose formats were designed around harvesting, transporting, and storing biomass as animal food (baled hay).

The second method as mentioned above is called cut stalk harvesting and is commonly used in Europe. In this method, there are three main techniques for harvesting hemp. In the first process, the hemp plant is taken into a singular knifed cutting drum and cut approximately 600 to 700 mm. The second commonly used process has two-barrel shaped cylinders with knives on them that cut the upright standing hemp. The third harvesting process is perhaps the simplest, with three offset four-meter bars vertically spaced apart by one meter. This last process utilizes a cutting head attachment to forage harvesters allowing harvesting of hemp to be more accessible to farmers, instead of having to buy a whole new machine. The three processing options are generally for short stem harvesting mentioned above are primarily used in industries that are not textile (Pari et al., 2017).

Most decortication technologies require that the straw not be cut too short. If cut to lengths on the order of inches, it is very difficult to maintain fiber length at the end of the decortication process. Storage is also critically important. Major industrial facilities will process between three to five tons raw material per hour. With only one harvest annually, the dry storage space required for a year's worth of bales would be dozens of acres.

Processing

Processing of hemp stalk for fiber has been performed for centuries. Though the technology has certainly changed (Fig. 6), the basic premise referred to as decortication remains the same; separate the bast fiber from the hurd, or the 'cleaning' process, and separate the bast fibers from each other, known as the 'opening' process. The largest economic impact will come from industrial cropping and processes, but there is likely to always be a sizable market around artisan fibers. The farm to consumer mentality has never been more pervasive than today and there is a strong hobbyist demographic utilizing the fiber in things like knitting. Artisan-scale processing may be performed with simple mechanical hand tools, while industrial processing will utilize large-scale, modern, automated equipment.

Fiber separation may be approached in several different ways, resulting in different fiber types each with unique opportunities for utilization. Applications for fibers exist at each step along the process shown in Fig. 7. Processes for the extraction of hemp fiber from the straw were developed from the linin and cotton industries. Modern flax straw processing starts with a breaking and scutching operation that physically separate the bast fiber from the hurd. Breaking usually involves the passing of parallel flax stems through crushing rollers to break the shive but not the fiber. With the shive and fiber loose, the fiber is combed with flat knives in a scutching tunnel. This process preserves the length of the flax fiber for long-line linen spinning operations. The short fiber that falls out of the scutching processes is referred to as scutched tow and is used in technical applications such as nonwovens for automotive applications. The design of flax equipment has been optimized around a standard flax straw length of around one meter. Hemp plants, which are usually greater than two meters in height, are typically not compatible with this process. Moreover, the major growth in usage of hemp bast fiber will be for use in technical or industrial applications, and not long-line spinning.

Decortication Process

The majority of fiber processing will come from commercial processing plants with throughput capacities in the tons of material per hour. These complex systems are purpose-built and engineered with robustness as a primary design requirement. The decortication process fundamentally consists of two steps; cleaning and opening. The goal of the cleaning process is the same as in flax systems; to separate the bast fiber from the inner hurd. This is differentiated from the opening process where the goal is to separate the bast fibers bundles from each other into smaller, finer bundles or filaments. The separated fiber bundles are typically referred to as "fibers", but that nomenclature is slightly misleading because as is shown in Fig. 7, these are actually bundles of fibers made up of elemental fibers.

Although decortication can technically be done by hand, as was done for hundreds of years, industrial scale systems are almost entirely mechanized. With current industrial systems, both the cleaning and opening steps, although described separately, are related and can be performed in the same processing line. That is, the equipment does not differentiate the fiber from the hurd, but rather the different

Fig. 6. Example of artisan processing versus industrial processing (Permission: University of Kentucky– Left, Sunstrand– Right).

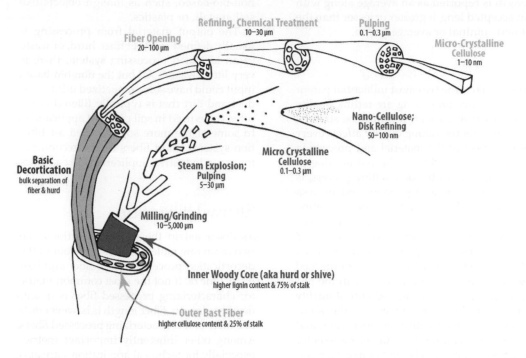

Fig. 7. Diagram correlating stalk morphology with processing technologies (Permission: BMI/Sunstrand).

The labels in the figure:

Refining, Chemical Treatment
10–30 μm

Pulping
0.1–0.3 μm

Micro-Crystalline Cellulose
1–10 nm

Fiber Opening
20–100 μm

Nano-Cellulose;
Disk Refining
50–100 nm

Micro Crystalline Cellulose
0.1–0.3 μm

Basic Decortication
bulk separation of fiber & hurd

Steam Explosion;
Pulping
5–30 μm

Milling/Grinding
10–5,000 μm

Inner Woody Core (aka hurd or shive)
higher lignin content & 75% of stalk

Outer Bast Fiber
higher cellulose content & 25% of stalk

materials respond differently to the mechanical processes. Furthermore, the steps are not mutually exclusive. Equipment designed to clean does some opening of the fibers, and equipment designed to open the fibers performs some cleaning. There is a balance that must be maintained between performing the processes to the degree required to yield high quality materials and overprocessing, which can result in unintended degradation of materials.

The mechanical separation of the fiber from the hurd in the cleaning step has the added benefit of reducing the hurd size and may negate the need for further processing of the hurd. On the other hand, the opening process is intended to reduce the fiber bundle diameters and runs the risk of causing unwanted degradation to the fibers. It is critical to understand this concept if natural fibers are to be competitive with replacing synthetics. The production of synthetic fibers is highly controlled, and the distribution and variance from nominal fiber length and diameter is generally very small. Due to the inherent nature of the variability of natural fibers, this is much more difficult to control.

Depending on the application and material specification, secondary processing of the fiber and/or hurd may be required. It is again important to understand that the goal of the processing is to produce consistent material, especially in technical applications where the natural fiber is replacing a highly engineered synthetic fiber.

Fiber length distribution can be difficult to control in high throughput decortication processing lines where the primary mechanism that reduces the fiber length is breaking and not cutting. A combination of fiber morphology, input quality (e.g., properly retted), and equipment design are among the variables that dictate the resultant fiber length. Secondary processes (e.g. cutting) that more accurately reduce the fiber length may be required at additional costs.

At present, there are no universally accepted guidelines for determining quality control metrics, or even how those metrics are determined. The best practices for quality control are a mix of similar yet borrowed techniques from other industries and relatively simple statistics. All raw and processed materials, whether synthetic or otherwise, exhibit some variability for any measurement and the inherent variability in the natural materials requires quality control (QC) and quality assurance (QA) documentation. Typically, processed fiber length is reported as an average along with an accepted length greater or lesser than the stated nominal or average length.

Factors Affecting Processing

In summary, the two most influential parameters on fiber processing are retting quality and moisture content. It cannot be stressed enough how the retting quality affects every other aspect of the material and processing outputs. The quality of the retting, due to the variability in the dew retting process, is largely qualitative. Agronomy and processing experts develop a "feel" for the retting quality and use identifiers such as apparent fiber strength, straw stiffness, color and other factors to help them form opinions about the retted quality. Moisture of the feed stock is the second biggest factor in the processing efficiency, yields, and output quality parameter. Moisture content of the straw can be quantified with moisture meters, and processors typically have upper bounds for what is acceptable. The higher moisture content reduces the efficiency of the processing equipment but also reduces the yield in two ways; the first in that moisture reduces the opening of the fibers and leads to high percentages of unopened fibers that are usually undesirable, and second, as the moisture content increases, the processor is paying the producer for water mass and not fiber mass.

The specific plant variety can influence processing in that certain varieties bred for fiber production can have higher fiber content and higher performing fibers. The harvesting equipment and harvesting techniques used mostly affect the throughputs in the processing equipment. Depending on the robustness or design of the front end of the line, very tightly packed or rolled bales can prevent challenges to the processing equipment. Plant populations can also dictate efficiencies and output quality. The plant density of the crop, affected by several parameters (e.g., seeding rate and success of establishment), can produce over-sized, under-sized, or highly variable straw that can be difficult for the processing equipment to handle effectively. Another consideration for output quality is the number of contaminants in the input straw. These contaminants may be bio-based, such as weeds or unwanted plant matter, or non-bio-based, such as foreign objects like soil, metals, or plastics.

The output material from processing is usually defined as either fiber, hurd, or waste. In an idealized processing system, there is very little "waste". All but the non-bio-based input could have some monetized utility. The dust and dirt that is typically filtered out is often times used in soil compost applications. In some of the more sophisticated air filtration systems, short fibers can be recaptured for use in industrial applications like plastics.

Straw Utilization

As discussed in the previous sections, the two main raw materials that come out of the decortication process are bast fiber and bast hurd. One of, if not the most common metric for characterizing processed fiber is quantified fiber length. Fiber length is however only one metric for characterizing processed fibers. Among other inherently important metrics, especially for technical applications, are also diameter, fibrillation, openness, color, and to lesser extent in most applications, smell.

Fiber diameter plays a crucial role in thermal and acoustical insulation performance. In order for natural fibers to compete or even be compared with ubiquitous synthetics, fiber diameter not only needs to be characterized, it needs to be controlled. Fiber diameter is related to the openness of the processed fibers, but there is a distinction between the two as well. The diameter of the fiber (bundles) is assumed and accepted to be approaching round or oval. Unopened fibers are described as "ribbonous" or "straps" and may only be one fiber (bundle) thick, but several bundles wide. Qualitatively, it is easy to identify ribbonous material; however, it is hard to quantify the diameter of fibers.

Another characteristic of processed fibers and the one that can lead to some complications in certain applications is the fibrillation of the fibers. Synthetic fibers tend to be smooth with consistent diameters. Natural fibers are not consistent and have varying diameters along their length, mostly due to discontinuous elemental fibers. These fiber ends can pull away from the fiber bundle and appear as tiny hair-like attachments. Both poor retting quality and over- processing can lead to the formation of fibrils.

Color may or may not be of concern in some applications, but in fact positive aesthetics of the fibers are often desirable. Consistency in color is often an indication of retting quality. Color is most important when some form of bleaching may be applied downstream.

Finally, the smell of the fibers may or may not be a concern for some applications. Automotive applications tend to be sensitive to smells in interiors, mostly from the off gassing of synthetics, but a "hay-like" smell from the fibers can sometimes be identified if the fibers are not fully encapsulated in a polymer matrix. A common application of a natural fibers encapsulated in a polymer matrix in automotive applications are the substrates that make up the head, trunk, and door liners where the natural fibers can constitute up to 50% of the mass of the panel.

Although not nearly as valuable as bast fibers on a per pound basis, the hurd produced from bast processing is also a very important for both application and economic reasons. Although the bast fibers are upwards of two to three times the value of the processed hurd, there is at least twice as much hurd in a bast plant as there is bast fiber on a mass basis, more often closer to 75/25 percent hurd to bast, respectively. Depending on the quality produced, the hurd could actually provide equivalent or more revenue than the fiber for each pound of input material. The simplest applications of the hurd require the least amount of processing and tend to command lower prices. These applications include animal bedding and hempcrete applications. More technical applications such as plastic fillers require further processing to make the hurd sizes and fraction more homogenous, in addition to creating fractions much smaller than what typically comes straight off a processing line. The three main attributes of the hurd that are usually of concern are particle size, aspect ratio, and color. Processed hurd sizes are typically reported as having an upper bound, or a percent content of different fractions, each having a unique distribution. Understanding the variances is critical not only for production analysis, but also for designers and engineers to fully realize the benefits of natural materials in their applications.

Green Plastics

In plastics, fibers are introduced to improve physical properties such as stiffness, impact resistance, bending and tensile strength. Synthetic fibers of glass, Kevlar, and carbon are most commonly used today, but plant fibers offer the potential for considerable cost savings along with achieving comparable performance. Natural fiber reinforced composites will not always meet the same strength properties as synthetic fibers, but the amount of fiber in a composite part can be increased to match the required stiffness of that part. The increase of fiber composition will decrease the amount of matrix material, polymers such as high-density polyethylene (HDPE), low density polyethylene (LDPE) and polypropylene (PP), which is generally a positive outcome. In fact, it is common in many plastic applications to offset some of the plastic with a lower cost filler, these fillers are often a mined mineral. The semi-crystalline nature of talc allows for the heightened compatibility with polymer matrix material with no loss of impact resistance. Fractionated hurd can easily be a competitor with materials such as talc. While typically not affecting tensile strength, use of hurd fillers can improve tensile and flexure stiffness, flexural and impact strength, and reduce the density of a plastic part. Properties such as these are indicative of hurd material due to the semialignment of cellulose chains throughout the particles.

Most composite car bumpers have talc introduced to reduce the weight of the bumper. The use of a small particulate size as a filler instead of the ubiquitous minerals can impart cost savings, increased material properties as well as renewable or recycling advantages. The energy consumption regarding these matrix materials are twice as much as the production of fiberglass. For some plastic parts in automobiles consisting of 100% polypropylene, natural materials can be introduced to replace some PP, maintaining strength requirements and producing a lighter part. In addition, compounders producing 3D-printing filament with polylactic acid, a corn-based biodegradable plastic, have begun adding micron-sized filler to enhance the biodegradable aspects of products produced by these methods.

In addition to lower production energy values, natural fibers have an average density of 1.4 g cm^{-3} while glass fiber has a density of 2.54 g cm^{-3}. Thus, even though natural fiber-polypropylene material requires a high volume fraction of natural fiber to achieve similar strength, the material will have a lower composite density of up to 40%, making the part lighter. Due to the lightweight nature of hemp fiber reinforced composite parts, weight advantages can increase fuel economy and reduce CO_2 emissions by reducing vehicle weight for mass transit and aviation as well as automotive sectors.

Industrial and Technical Textiles

The exact types of fibers produced in the decortication process are largely application dependent. Longer fibers that are greater than 1.5 inches in length tend to see application in higher value, technical applications. Shorter fibers are utilized in applications where the specifications are not as strict as the longer fiber applications and tend to be of lower value. Whereas longer fibers are utilized based on their highly specific properties such as stiffness to weight ratios, short fibers may tend to be utilized in less value driven applications such as low strength fillers, as opposed to the hurd fillers whose value is in light weighting.

The bast fiber from the hemp plant is generally compatible with typical nonwoven manufacturing technologies; carding, air-lay, and wet-lay. However, most of the commercial nonwoven systems in place have been designed to run either synthetic fibers or cotton. Variability in fiber length and increased dust or fines can pose challenges for manufacturers to utilize hemp fiber. Nonwoven lines designed to run cotton shoddy or other natural fibers will typically have air filtration and machine parameters (wire or clothing) which are more conducive to hemp fiber.

There are a wide variety of applications for hemp fiber nonwovens. Thermoformable mats can be made for applications like automotive door panels. Some air filtration media and geotextiles could utilize hemp fiber. Hemp fiber has good acoustic and thermal insulation properties, and therefore, will likely see good success as acoustic insulation (commercial construction, appliances) and thermal insulation (building insulation, insulated packaging, etc.). Thin nonwoven webs, typically referred to as veils or scrims in composite application, could utilize hemp fiber.

Garment and Home Furnishings Textiles

Textile and spinning applications, depending on their downstream processing technologies can utilize both short and long fibers. These applications typically require fibers that are delignified and reduced fiber bundle diameters. Often a second step is required using chemical treatments to further process the fibers to clean, bleach and to reduce fineness. Chemical delignification or degumming may also be required to aid in the mechanical processing as well as allow for better bleaching and dying.

Hemp fiber blended with cotton has recently seen some success in workwear application where it is has been shown to increase durability and reduce weight compared to 100% cotton counterparts (Patagonia, 2019) High-grade hemp fiber can be compatible with yarn spinning; however, there is significant cost to the fiber processing. Fabrics used in home furnishings (floor coverings, upholstery, draperies, etc.) generally have a lower grade of "hand" than garments; therefore, the requirements on hemp fiber are stringent, enabling the utilization of courser and less clean fiber.

Absorbents (Industrial and Animal Bedding)

Common uses for hurd are generally applications that require no to little extra processing on the manufacturer's end. The simplest applications of the hurd require the least amount of processing but also tend to command lower prices. Low value uses of hurd, such as animal bedding and spill kits, are in high demand due to the high absorbency nature of the hurd material. Hemp hurd is generally considered to be a high-grade animal bedding option. If processed correctly, hemp hurd will have low dust, high moisture absorption, and require less stall and cage turnovers compared with competitive products on the market.

Construction Materials (Hempcrete, Hardboard, and Insulation)

Hempcrete is a hurd and lime mixture creating a non-loadbearing, breathable wall material for construction applications. While still considered a niche wall-construction technology, it is gaining popularity. Typically, a building is constructed with a standard wood framing system. The hempcrete mixture is then used to infill the studs eliminating the need for other insulations. Many installers will coat the hempcrete with an adobe or plaster finish on both the inside and outside. If installed properly, this can in same case eliminate the need for a moisture barrier and exterior siding.

Another use of large format hurd with similar specifications to hempcrete is in fiber board applications. Medium density fiberboard manufacturers have pursued the use of hurd as a replacement for wood. However, market studies indicate that hurd is not yet price-competitive with wood, and they do not yet meet mechanical performance standards needed for structural sheeting applications in building construction. Boards with more aesthetic end uses such as ceiling tiles and breathable partition walls for office spaces are, however, viable options (Vuorinen, 2019; Strunk, 2012).

The unique structure and morphology of the fibers are of interest for thermal and acoustic insulation applications. The surface of the fibers tends to not be as smooth as synthetic fibers and it is thought that this surface roughness seems to increase insulation properties by inhibiting airflow and redirecting sound waves. Early studies and product development have shown that in certain configurations, natural fibers can meet and sometimes exceed traditional synthetic fiber performance in insulation applications.

Pulping

Generally, the main purpose of pulping is to completely remove the lignin matrix from the cellulose fiber in a lignocellulosic biomass. Cellulose fibers are used in a variety of industrial, pharmaceutical, and food applications. There are a variety of cellulose derivatives on the market that are used in multiple sectors and industries. However, all cellulose products and derivatives start with pulp. The pulp is formed either through a chemical or mechanical method and is the precursor to paper and paperboard products. Specialty pulps can be made via different production methods that have other end uses aside from paper and paperboard products.

While pulped cellulose fiber from both hemp bast fiber and hurd can be compatible with papermaking, the cost is presently prohibitive compared with wood-based pulps.

In general, cellulose consists of amorphous regions, which can be treated with a strong acid and produce microcrystalline cellulose (MCC) or nanocrystalline cellulose (NCC) (Vuorinen, 2019; Fuqua et al., 2012). Microcrystalline cellulose is used widely in the food and pharmaceutical industry. Nanocrystalline cellulose has a wide variety of potential applications such as food packaging, aerogels, thin films, and composites. Recent research efforts indicate that lignin could be a potential biological source for many important chemicals.

One application at the forefront of the composite industry is nanotechnology. Cheap sources of carbon nanotubes are now emerging on market; bulk sales of nanocellulose are also increasing at a pilot scale, but still costly. The single chains of cellulose can be created from breaking down cellulose-based material. The polysaccharide chains form into either nanocrystals or nanofibrils and both can be used for composite applications. The use of nanocellulose for composite applications is currently deterred due to its hydrophilic tendencies (Mossello et al., 2010; Strunk, 2012; Mossello et al., 2010; Sain et al., 2002).

The wood pulping industry is very established, and since the lignin structures in wood are different from the lignin structures in bast plants, processing variations are necessary to use bast-sourced lignins. Unfortunately, there is not a lot of information about hemp or other bast plants in regard to pulping. However, hemp and other bast fiber plants are renewable crops, and pulping of these plants is more environmentally friendly. Compared with wood, hemp shive has relatively high lignin contents, so delignification can be difficult (Chandra, 1998).

Advanced Technical Applications

Advanced applications must be investigated, such as activated carbon. Production of activated carbon substrates require carbon rich materials that need to undergo pyrolysis. Currently, the main source of activated carbon is coir fiber from the coconut husk but the global supply is currently decreasing. Torrefied hurd can be used as filler in the same manner (Ibrahim et al., 2010; Solfa et al. 2016;

Nsor-Atindana et al., 2016). This material has been shown to withstand pyrolysis and eventually used to produce a supercapacitor, an entity currently seen as the potential future of batteries (Karimi et al., 2014; Kian et al., 2017; Prosenjit et al., 2016; Sun et al., 2016). For global electrical networks to survive in the future, a biodegradable battery that is capable of fast charging and slow discharging is required. This would make existing renewable harvesting technologies such as solar and wind more efficient. For a less resource intensive product development process, investigating the use of hemp-derived biochar as a supplement for agricultural fields can be investigated. Spreading both non-pyrolyzed and pyrolyzed material for field management has been shown to increase nutrient content in fields (Chandra, 1998; Ardanuy et al., 2015; Krotov, 1995).

Economics

As with all industries, but particularly emerging ones looking to display new but currently accepted products, economics are the cornerstone. Fundamentally, hemp bast fiber and hurd products must compete with many existing materials on a cost and performance basis to be successfully adopted in any meaningful application. These price points and performance metrics are already set in industry. Hemp value is precisely tied to that of the incumbent material. The bast fiber must compete on a value basis against a wide range of fiber from synthetics (e.g. polyester, polypropylene, glass) and other natural fibers (e.g. cotton, jute, animal). Similarly, the hurd must generally compete with other short fiber and particulate materials (for example, wood, talc, and silica). While advancements into exotic material application are possible, driving the value of the hemp straw higher, the near and intermediate applications will be based on utilization in existing commodities.

The product price will then drive allowable costing through the hemp fiber supply chain. Hemp bast fiber and hurd sales must support the operational costs and standard profitability; however, early investments in infrastructure may be hampered by limited historical market adoption, higher costs of goods prior to industrial economies of scale at the process and producer level, as well as increased requirements on businesses for working capital to support process, product, and market development. Considering the high cost of infrastructure for fiber processing applications, limited operational and market data deployment of a capital into the hemp industry will likely be viewed as higher risk. Therefore, applications which garner higher price points and subsequently higher margins are the likely early targets. At the farm level, cropping for hemp straw will compete with major row commodity crops, such as corn, soybeans, and canola.

Future and Standardization

Though hemp fiber is not new, its place in industry today is insignificant compared with other fibers. For the hemp fiber industry to grow, it will require an evolution in which industry stakeholders unite to converge on terminology, identify deficiencies in the value chain, and develop universal specifications wherein grading is understood in the context of compatible applications and value. Trade organizations will play a crucial role in supporting cooperation among industry members to achieve these goals. One benchmark example to follow would be the cotton industry. For many years cotton has had the support of both government regulators and industry associations. Cotton, Inc., the main cotton trade organization, articulates very well an approach to standardization which includes (Cotton Inc., 2018):

- The purpose of standards is to create a universal system for measuring fiber and product quality. Standards are a business tool, and in many cases, standards are a strategic step for developing new global markets. Standards ensure trade by eliminating trade barriers, saving companies money, and by accelerating research.

- There are standard ways that value is measured and assessed to products. Many inspection, certification and approval systems are based on global standards.

- Measurement of these values supports the ability to grade and value the fiber.

- Methodologies for characterization must be agreed upon and employed by all parties for the system to be meaningful.

As noted, fibers have value and may be utilized in different applications at each point in

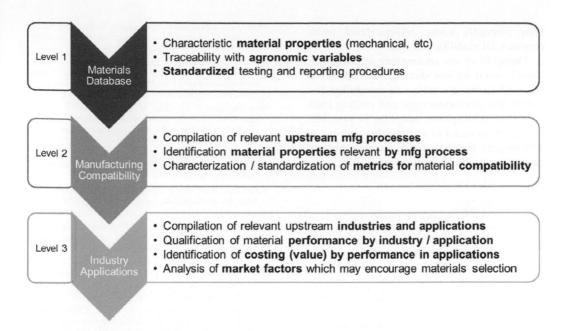

Level 1	Materials Database	• Characteristic **material properties** (mechanical, etc) • Traceability with **agronomic variables** • **Standardized** testing and reporting procedures
Level 2	Manufacturing Compatibility	• Compilation of relevant **upstream mfg processes** • Identification **material properties** relevant **by mfg process** • Characterization / standardization of **metrics for** material **compatibility**
Level 3	Industry Applications	• Compilation of relevant upstream **industries and applications** • Qualification of material **performance by industry / application** • Identification of **costing (value) by performance in applications** • Analysis of **market factors** which may encourage materials selection

Fig. 8. Levels of understanding and efforts required for advancing the utilization of natural fibers in modern, technical applications (Permission: BMI/Sunstrand).

the processing diagram shown above in Fig. 7. To ensure that processing results in end-use application, characterization must follow as outlined in the three-level diagram below (Fig. 8). The depth and breadth of potential applications is immense as access to these materials that have heretofore been predominantly unavailable or under-engineered. Without proper material characterization and critical value assessment, these opportunities will be lost or perhaps never even realized. In short, the success of hemp fiber for industrial and technical applications is entirely a function of performance. Through the development and use of standardized testing and grading procedures at both the material and manufacturing levels, identification and qualification for industrial applications must be achieved.

The most common method of testing to assess the effects of these variables and technical applications is a tensile test performed in accordance with the ASTM standard C1557. This defines the standard testing methods for both the tensile strength and the Young's Modulus of fibers (ASTM, 2014). To prepare a sample for this test, one must mount the fiber onto a tab with a window cut out revealing the section of fiber to be tested. Ensuring the fiber remains straight and parallel to the testing direction is not trivial.

In apparel applications, industry has developed around the metrics that are necessary to ensure proper spinning of the fiber. Spinning technologies were developed in concert with the material properties primarily of cotton and synthetic thermoplastic fibers.

While test standards exist for natural fibers, they are typically embedded in, or are variations of, synthetic fiber test methods. However, existing standards do not consider the variation in the natural fibers, resulting in potentially inaccurate definitions of material properties. To help overcome this and develop the material database required to realize significant material implementation, we must develop standards to assess and examine the effect of these variables. The results from these trials will provide data on the commercial viability of processing bast fiber for technical applications. It is imperative to find the balance between the ability to process the bast while preserving the required performance metrics. In addition, the impact of each variable on fiber fineness, fibrillation, strength, and stiffness will allow for clear identification of the key innovative material processing methods, as well as the most appropriate materials applications. This work will for identification of not only what hemp fiber is, but also what it is not, validating or refuting claims around topics such as hemp fiber strength, costs, anti-microbial traits, commercial viability, et cetera.

Hemp fiber has an amazing opportunity ahead given its low density and favorable strength-to-weight ratio and durability. It's versatility, performancgge and costing indicate that widespread adoption is possible. However, in order to achieve this success, it will require the industry to focus on proper characterization and education.

References

Ardanuy, M., J. Claramunt, and R. Filho. 2015. Cellulosic fiber reinforced cement-based composites: A review of recent research. Constr. Build. Mater 79:115–128. doi:10.1016/j.conbuildmat.2015.01.035

Asby, R. 2009. Syst. Pract. Action Res. 22:357

ASTM. 2014. ASTM C1557-14, Standard test method for tensile strength and young's modulus of fibers, ASTM International, West Conshohocken, PA. doi: 10.1520/C1557-14

Chandra, M. 1998. Use of nonwood plant fibers for pulp and paper industry in Asia: Potential in China. VTechWorks. Virginia Tech, Blacksburg, VA.

Cothren, J. 2014. Advantage of crop rotation. Wilkes County Center; NC State University Cooperative Extension, Wilkesboro, NC.

Cotton Inc. 2018. Classification of cotton. Cotton Inc., Cary, NC. https://www.cottoninc.com/cotton-production/quality/classification-of-cotton/. (Accessed 17 August 2018). [2018 is year accessed].

Dai, Q., J. Kelly, J. Sullivan, and A. Elgowainy. 2015. Life-cycle analysis update of glass and glass fiber for the GREET model. Systems Assessment Group; Argonne National Laboratory, Lemont, IL.

Fuqua, M.A., S. Huo, and C.A. Ulven. 2012. Natural fiber reinforced composites. Polym. Rev. (Phila. Pa.) 52(3):259–320. doi:10.1080/15583724.2012.705409

Golden, J.S., R.B. Handfield, J. Daystar, T.E. McConnell. 2015. An economic impact analysis of the U.S. biobased industry. USDA, Washington, D.C.

Government of Alberta. 2004. Beneficial management practices: Environmental manual for crop producers in Alberta- Cropping rotation. Government of Alberta, Agriculture and Forestry Division, Edmonton, AB.

Hockstad, L., and M. Weitz. 2015. Inventory of U.S. greenhouse gas emissions and sinks: 1990-2014. Environmental Protection Agency, Washington, D.C.

Ibrahim, M., F. Agblevor, W. El-Zawawy. 2010. Isolation and characterization of cellulose and lignin from steam-exploded lignocellulosic biomass. BioResources 5(1)397–418.

J.E.C. Composites Publications. 2014. Flax and hemp fibres: A natural solution for the composite industry. JEC Group, Paris.

Karimi, S., P.M.D. Tahir, A. Karimi, A. Duresne, and A. Abdukhani. 2014. Kenaf bast cellulosic fibers hierarchy: A comprehensive approach from micro to nano. Carbohydr. Polym. 101: 878–885. doi:10.1016/j.carbpol.2013.09.106

Kian, L., M. Jawaid, H. Ariffin, and O. Alothman. 2017. Isolation and characterization of microcrystalline cellulose from roselle fibers. Int. J. Biol. Macromol. 103: 931–940. doi:10.1016/j.ijbiomac.2017.05.135

Krotov, V S. 1995. Use of AAS pulping for flax and hemp shives. Ukrainian Pulp and Paper Research Institute, Kiev, Ukraine.

Mossello, A., J. Harun, H. Resalati, R. Ibrahim, P.M. Tahir, S.R.F. Shamsi, and A.Z. Mohamed. 2010. Soda-anthraquinone pulp from Malaysian cultivated kenaf for linerboard production. BioResources 5(3):1542–1553.

Nsor-Atindana, J., M. Chen, H. Douglas Goff, F. Zhong, and Y. Li. 2017. Functionality and nutritional aspects of microcrystalline cellulose. Carbohydr. Polym. 172: 159–174. doi:10.1016/j.carbpol.2017.04.021

Pari, L., P. Baraniecki, R. Kaniewski, and A. Scarfone. 2017. Harvesting strategies of bast fiber crops in Europe and in China. Ind. Crops Prod. 68: 90–96.

Patagonia. 2019. Hemp fabric. Patagonia, Ventura, CA. https://www.patagonia.com/hemp.html [2019 is year accessed].

Penn State College of Agricultural Sciences Cooperative Extension. 1996. Crop rotations and conservation tillage. Penn State, University Park, PA.

Prosenjit, S., S. Chowdhury, D. Roy, B. Adhikari, J.K. Kim, and S. Thomas. 2016. A brief review on the chemical modifications of lignocellulosic fibers for durable engineering composites. Polym. Bull. 73(2)587–620.

Sain, M; Fortier, D; Lampron, E. 2002. Chemi-refiner mechanical pulping of flax shives: Refining energy and fiber properties. Bioresour. Technol. 81: 193–200.

Solfa, M.R.K., R.J. Brown, T. Tsuzuki, and T.J. Rainey. 2016. A comparison of cellulose nanocrystals and cellulose nanofibers extracted from bagasse using acid and ball milling methods. Adv. Nat. Sci.: Nanosci. Nanotechnol. 7(3):035004.

Strunk, Peter. 2012. Characterization of cellulose pulps and the influence of their properties on the process and production of viscose and cellulose ethers. Umea University Department of Chemistry, Umea, Sweden.

Sun, W., S. Lipka, C. Schwartz, D. Williams, and F. Yang. 2016. Hemp-derived activated carbons for supercapacitors. Carbon 103:181–192. doi:10.1016/j.carbon.2016.02.090

U.S. Environmental Protection Agency. 2015. Documentation for greenhouse gas emission and energy factors used in the waste reduction model (WARM). Environmental Protection Agency, Washington, D.C.

Vosper, J. 2011. The role of industrial hemp in carbon farming. GoodEarth Resources PTY Ltd., Sydney, Australia.

Vuorinen, T. 2019. Chemistry of pulping and bleaching. Aalto University School of Chemical Technology, Helsinki, Finland. [2019 is year accessed].

Williams, D.W., J. Wade Turner, R. Hounshell, D. Neace, and T. Riddle. 2017a. The effect of seeding rates on yield and stem growth parameters of kenaf (Hibiscus cannabinus L.). Unpublished.

Williams, D.W., J. Wade Turner, R. Hounshell, D. Neace, and T. Riddle. 2017b. The effects of seeding rates and row spacing on yield and stem growth parameters of hemp (Cannabis sativa L.). Unpublished.

Photo by University of Kentucky Department of Agronomy

Chapter 4: Hemp Agronomy-Grain and Fiber Production

Jeff Kostuik and D.W Williams*

Introduction

Much of what is provided in this chapter is gathered from ongoing applied research trials in the Parkland region of Manitoba, beginning in Canada from 1998 when hemp production was once again legalized and commercial production recommenced. When industrial hemp was legalized in Canada in 1998, past literature shows that much of the early research on hemp agronomy was specific to growing hemp for fiber. As early as 1920, Agriculture Canada was involved with hemp research as a part of the fiber crops program. Additional information in this chapter is derived from research work at the University of Kentucky during the 2015 through 2018 growing seasons. Lastly, we are also reporting on common-knowledge agronomic factors affecting hemp establishment, culture, and harvest.

Hemp is both a heliotrope (sun-loving) and thermophilic (warmth-loving) crop, (Clarke and Merlin, 2013). When conditions are right, it produces large amounts of biomass, pollen, and seed. It is not terribly difficult to grow hemp as a crop, but it is certainly possible to mismanage the establishment and/or crop culture phases such that failure is the outcome.

One of the main ongoing issues associated with hemp production continues to be stand establishment. There are some key environmental factors that influence the early growth and development of hemp. These include sunlight, temperature, soil condition or tilth, and soil moisture. However, research has shown that for many of these environmental factors, there are agronomic factors that affect the probability of a poor stand. These include field selection, variety selection, seed bed preparation, seeding depth, fertility rates, fertility placement, and likely as important as any, seeding date.

The second and arguably equally important issue with hemp production is growing and maintaining a high-quality product. Hemp quality starts in the field! This is true for both fiber and grain production. There is very little processing done to a hemp seed prior to finding its way onto the consumer's plate, with little in the way of kill stops such as chemicals or heat used to control microorganisms like bacteria, molds, and yeast. Therefore, producers must be vigilant when growing, handling and maintaining a quality product throughout the growing,

J. Kostuik, Hemp Genetics International, Langley, BC V1M3V5. *Corresponding author (dwilliam@uky.edu)

doi:10.2134/industrialhemp.c4
© ASA, CSSA, and SSSA, 5585 Guilford Road, Madison, WI 53711, USA.
Industrial Hemp as a Modern Commodity Crop. D.W. Williams, editor.

Table 1. Nutrient Uptake and Removal by Hemp and Canola (kg ha⁻¹)

Nutrient	Total Plant Uptake (Kkg ha⁻¹)		Removal by Grain (kg ha⁻¹)		Total Nutrient Uptake per day (kg ha⁻¹)
	Hemp*	Canola**	Hemp*	Canola*	Hemp**
Nitrogen	200	120	40	65	6.7
Phosphate	47	50	19	35	1.6
Potassium	211	75	10	17	6.0
Sulfur	14	20	3	12	

† Source: Canadian Fertilizer Institute, 2001

‡ Source: Manitoba Agriculture, Food, and Rural Development, 2019

harvest, drying and storage aspects of hempseed grain grown for food. Proper retting of hemp fiber will define the quality of the final product regardless of many prior management decisions. Additionally, proper storage of baled hemp fiber until processing is imperative to maintain quality.

Variety Selection

Choosing the best variety for a specific environment, end use (s), and geographic location is very important in the potential success of a hemp crop. Local data for hemp varieties is still somewhat scarce in most regions. However, it is paramount that varieties be tested in various locations under different climate conditions to ensure producers have the most accurate assessment of local variety performance. This certainly includes adaptation to latitude.

It can be stated that hemp breeding and varietal development is in its infancy in North America so there are not many factors, other than seed size and plant height that set one variety aside from another. Variety testing in Canada has looked at both dual-purpose and grain-only varieties. Some work at the University of Kentucky has evaluated fiber-only varieties.

With dual-purpose varieties, the producer harvests the seed for grain and then harvests the mature fiber after grain harvest. These taller high yielding grain varieties may prove challenging for grain harvest as you must push much more biomass through your harvesting equipment.

There is a steady movement toward breeding for shorter-statured, high-yielding grain varieties for grain production only.

Harvestability is the key with these varieties and they are recommended for first time hemp growers looking to harvest seed for grain as they will have less challenges with their equipment during harvest.

The take home message is to know what market the end-product is going to. This will dictate the proper agronomic advice, variety selection, and the plan for harvest, drying, and storage.

Field Selection
Soil Properties

Hemp can grow and survive on a wide range of soil types. However, to provide high grain yields, hemp requires deep, well-drained soils because the plants cannot tolerate standing water (anaerobic conditions). Soils that have adequate tilth so as to not negatively affect plant available water, such as loam to sandy loam soils high in fertility are most recommended for hemp grain and fiber production. Soil characteristics such as salinity, compaction, and high acidity or basicity should be avoided. Hemp tends to perform best on soils with a pH range of 6.5 to 7.0, or slightly acid to neutral.

The following photo shows the effect of a three-inch rain following seeding resulting in drowned out area of the field. These areas provide a refuge for volunteer crops, and weed growth.

Crop Rotation

Hemp is an oilseed, and thus works best in a crop rotation with either a cereal crop or preferably a legume, just as any other field crop rotation. Again, like any other agricultural crop, hemp planted after hemp should be avoided for many reasons including potential increases in disease pressure, reduced fertility, and reduced food quality issues. For short-statured hemp varieties grown for grain, wheat should be avoided as a crop prior to hemp. Processed hemp grain is sold as a gluten free product and wheat can seldom be completely mechanically cleaned from the sample and thus the gluten free label cannot be used. Hemp is a host to *Sclerotinia* (white mold) and *Botrytis* (gray mold), which create food quality issues as well. Hemp grown as fiber can follow any previous crop. To date, there have been very few fungal diseases affecting the yields or quality of hemp fiber crops.

Fertility

Hemp performs very well on highly fertile soils with good soil structure that allows for good drainage and adequate root development. The following table represents a nutrient uptake trial conducted in Manitoba Agriculture in 1999.

Nitrogen

Nitrogen is a key nutrient in hemp grain production. Nitrogen is essential for chlorophyll, protein synthesis, photosynthesis, amino acids, utilization of sunlight, nutrient uptake, and energy systems. The key factors affecting nitrogen in the soil are the sources (fertilizer, previously legume crops, soil organic N), and the environment (temperature, humidity, and precipitation).

As seen in Table 1, hemp uptake of nitrogen represents a substantial contribution to hemp growth. With such a nitrogen-hungry plant, focus should be on the basic nitrogen fertilizer best management practices for optimal plant growth. The 4 R's of fertilizer best management practices are the Right source, the Right placement, the Right rate and the Right timing. Hemp has low tolerance to seed-placed fertilizer so it must be either banded prior to seeding or precision placed away from the seed at time of planting. Nitrogen recommendations for total N available (soil N plus additional added N) range from 100 to 150 lb acre^{-1} (112 to 168 kg ha^{-1}) for dry land and up to 200 lb acre^{-1} (224 kg ha^{-1}) for irrigated hemp grain crops. Timing of N requirement and availability

during the growing season are areas that could use some further research. As hemp moves into its elongation phase in early July, we have documented up to three inches of growth per day. During this time N uptake can reach levels higher than 6 to 7 lb per acre per day. Crop fertilization or top dressing studies are needed to determine if hemp will convert this extra N to seed production or to more biomass production.

Only 40 lb (18 kg) of the 200 lb (91 kg) of N are removed from the field as harvested hemp grain (average yield of a 1300 lb grain per acre [1457 kg ha^{-1}]). After grain harvest, the hemp fiber or straw sits or rets in the field. When the straw is left in the field, much of this mobile N may be leached back into the soil and become available for future crops.

The same level of heavy nitrogen fertilization of hemp fiber crops is not desirable. As noted above, extremely rapid growth rates occur during elongation. This type of growth in plants is mainly a function of cell elongation, which is exacerbated by applications of nitrogen fertilizer or generally high amounts of plant available N. High rates of cell elongation lead to thinner cell walls. Thin cell walls lead to weaker bast fibers. For this reason, only 50 lb N acre^{-1} (56 kg ha^{-1}) are recommended for hemp fiber crops, and this is usually applied pre-plant.

Phosphorus

After nitrogen, phosphorus generally has the greatest effect on hemp grain yield. Phosphorus is used by plants for photosynthesis, respiration, cell division, seed growth and development, seed formation, and most importantly, for ensuring vigorous early root growth.

Phosphorus availability is limited by soil moisture status, soil temperature, soil pH, and soil texture. Low soil moisture reduces P diffusion required for plant roots to access available P, and high soil moisture limits the availability of oxygen. Low soil temperatures decrease P availability by reducing the rate of root growth leading to nutrient absorption. Heavy textured and higher pH soils will "tie up" P reducing the amounts available to the growing crop.

Studies at Parkland Crop Diversification Foundation have shown that hemp is relatively sensitive to seed placed P. Applications of P_2O_5 over 20 lb actual P resulted in a significant increase in hemp seed mortality.

Table 2. Macronutrient needs for hemp crops.

	Fiber	Grain/dual purpose
Available Nitrogen	50 lb acre^{-1}	100–200 lb acre^{-1}
Available Phosphorus †	60 lb acre^{-1}	50 lb acre^{-1}
Available Potassium †	300 lb acre^{-1}	300 lb acre^{-1}

† Recommended phosphorus and potassium application rates are provided in Table 3 below and were derived from AGR-1 (University of Kentucky Agricultural Extension, 2018).

Table 3. Phosphorus and potassium application rates as a function of soil test results.

Category	Test result: P	P$_2$O$_5$ needed	Test result: K	K$_2$O needed
Very high			>420	0
High	>60	0	355-420	0
			336-354	0
			318-335	0
			301-317	0
Medium	46-60	30	282-300	30
	41-45	40	264-281	30
	37-40	50	242-263	30
	33-36	60	226-241	40
	28-32	70	209-225	50
			191-208	60
Low	23-27	80	173-190	70
	19-22	90	155-172	80
	14-18	100	136-154	90
	9-13	110	118-135	100
	6-8	120	100-117	110
Very low	1-5	200	<100	120

However, hemp has the ability to overcome the effects of increased seedling mortality through increased branching making up for lost plants and maintaining yield. At the same time, days to maturity will increase with thin plant stands. Therefore, hemp recommendations for seed placed P are always reliant on moisture, soil properties, seeder, or planter openers and temperature. On a clay loam soil with narrow openers (1 inch or less) no more than 20 lb of actual P$_2$O$_5$ with the seed is recommended. As P is an immobile nutrient, this will allow for early access for root development. Higher rates of P (if needed) can be banded or broadcast applied to meet long term crop needs. Fifty pounds of available P$_2$O$_5$ soil and applied fertilizer is recommended for a hempseed grain crop. Sixty pounds is recommended for fiber crops.

Potassium

Potassium is often referred to as the third nutrient. Potassium is essential for plants and its role includes, but is not limited to, kernel weight, root growth, grain-filling period, maximizing kernel weight, stress tolerance, disease and environmental stress resistance, and most importantly, stem strength.

It was no wonder that in nutrient uptake trials, the required soil K levels were extremely high, given the long stem associated with hemp. Like nitrogen though, little K leaves the field in the form of grain. Most of it is left in the field, and since K is somewhat mobile, if the stalks are left to ret in the field for some time, much of the K will be leached back into the soil. Producers must note that if the fiber is baled green or the total plant biomass is removed from the field prior to the retting process, large amounts of nutrients, including K will be removed. Nutrient removal will need to be addressed in some manner to maintain soil nutrient status over an extended period. Potassium deficiencies are known to occur in light-textured soils, peat soils, high rainfall areas, and intense cropping systems.

Weed Control

Weed control is extremely important for successfully growing hemp for grain and fiber and for maintaining quality of the harvested products. There are currently very few herbicides registered for industrial hemp even for minor use in Canada where the crop has been grown for more than two decades. No herbicides are registered for use in the United States. Without available herbicide options, producers must look toward mechanical and cultural control of weeds and take advantage of hemp's ability to out-compete weeds when properly established.

Establishment of an adequate plant population is the best strategy for weed control in hemp. Seeding shallow (0.5 to 1 inch), into a warm, firm, and moist seed bed to help hemp germinate and grow rapidly is crucial. Seeding rates of 25 to 30 lb acre^{-1} for a grain crop will produce a plant stand of 9 to 13 plants ft^2. Seeding rates for fiber crops should be 40 to 60 lb acre^{-1}. When seeding into less-than-ideal soil conditions that will increase plant mortality, higher seeding rates may be necessary. Hemp will be vulnerable to weed competition from seeding until full canopy cover, which will generally happen in two to four weeks after seeding. Once hemp is around six to eight inches tall, the aggressive growth and biomass production of the hemp plants will outcompete most weeds. There are possible exceptions to this. For example, Johnsongrass (*Sorghum halepense* [L.] Pers) and morning glory (*Ipomopea* spp.) are two weeds that may cause serious issues even after hemp canopy closure. Johnsongrass can easily grow just as rapidly and as tall as many hemp varieties, thus contaminating the crop. Morning glories are vining weeds that can literally grow over and cover shorter hemp crops and/or entwine among the stems of taller hemp crops. In either case, both the quality of the final product and the harvestability of the crops are negatively impacted. Research is underway in the United States to investigate herbicide usage in hemp crops for both monocot and dicot weeds.

Crop rotation and field weed history should play a large part of the field management and decision process when considering where to plant hemp. It is best to absolutely avoid fields with hard-to-control weeds. Proper fertility levels and nutrient placement is also key, as it will provide hemp the ability to establish and grow quickly and thus out-compete weeds. Using mechanical weed control such as a cultivation just prior to seeding will help give hemp the advantage over late germinating weeds. Interrow cultivation can be an option with wider row spacing (16–20 inches). However, row spacing greater than 16 inches inhibits the ability of most hemp varieties to provide proper canopy cover allowing weeds to continue to compete.

Regardless of practice, the importance of weed control in hemp production systems cannot ever be over emphasized. Contrary to some reports, hemp will not eliminate or reduce weed pressure in a field. Weeds alone are one of the most common causes of hemp crop failures in the United States. Additionally, seeds of many weeds and volunteer hemp crops are difficult to clean properly from hemp seed. There is an extremely low tolerance for foreign weed or volunteer crop seed in quality hemp for grain, and even less in certified seed for propagation of next year's crop. In Canada, hemp grain for dehulling is cleaned to a 99.95% purity specification. This is possible but cannot be guaranteed through seed cleaning systems that use wind and screens when weed seeds are present. Gravity tables, indents, destoners, and most importantly, color sorters are required to maintain the high-quality standards consumers expect and certified seed specifications require. Significant weed populations in baled hemp fiber (e.g., Johnsongrass) may be cause for rejection of the material at the processing plant.

Seeding and Planting

Hemp is a day-length sensitive crop and flowering occurs close to the same calendar date every year independent of time of seeding. Factors including fertility stress, weed pressure, and moisture will affect the precise flowering time. Finding the optimum seeding date for each variety providing for

Table 4. General recommendations for seeding rates and row spacings for dedicated hemp fiber crops and grain or dual-purpose hemp crops.

	Fiber	Grain/dual purpose
Seeding Rate (lb acre⁻¹)	40–60	20–40
Row spacing (in)	8	8–16

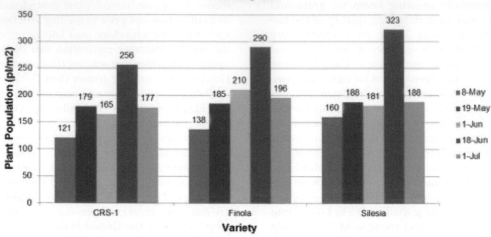

Fig. 1. Plant populations (plants per m²) as a function of planting date and variety.

best stand establishment and highest yield of grain or seed is probably still the most important research needed when establishing hemp as a viable crop in a new region.

Hemp requires relatively small amounts of water or moisture just to survive, but during critical times of the crop's life, such as the germination and establishment phase, adequate moisture is critical. Planted in warm soils (> 50 °F) and in the presence of adequate moisture, most hemp varieties will germinate in 3 to 5 d. In general, we classify hemp seedling vigor as relatively low compared with other crop species. Young, newly germinated hemp plants are susceptible to many potentially fatal stresses from biotic, physical, and environmental sources. Thus, planting date and seed bed preparation are much more important in hemp compared with almost any other traditional agriculture crop.

Also related to low seedling vigor, it can be important to consider the predicted weather soon after planting. This is especially critical under conventional tillage. If heavy rains exceeding soil percolation rates occur after seeding and before emergence

under conventional tillage, the smallest soil particles on the surface will become suspended within pockets of standing water. As the water drains or evaporates, the small soil particles will be deposited as a layer on the surface. Once dry, this layer may become quite hard and completely impenetrable to small hemp seedlings just germinated. This is often referred to as crusting. Crusting is rarely an issue over an entire field, but significant sections of fields may be severely affected, allowing for heavy weed establishment and crop failures in crusted areas. Avoid crusting by being aware of potentially heavy rains post-planting and before emergence, or by planting with no-till equipment/ protocols. Additionally, the same precipitation event occurring soon after emergence will likely drown hemp seedlings if drainage/evaporation doesn't occur within 48 h.

Seeding Rate

Hemp seed is expensive to plant. One must use the best methods possible to ensure you plant enough seeds to have an adequate stand without seeding too much or too little. Using the thousand kernel weight for seed rate calculation is a good way to determine your target seeding rate:

(lb/ac) = desired plant population/ft² ´ 1000 K seed weight. (g) ÷ seedling survival rate (in decimal form such as 0.90) ÷ 10.4

The desired plant population for hemp is 7 to 13 plant ft² in conventional grain, dual-purpose, and fiber production, and 10 to 13 plant ft² for organic grain production. The 1000 k wt (g) is the total grams that 1000 kernels of a variety of hemp weighs. This range is from 14 g for small seeded hemp up to over 20 g for large seeded hemp.

Seeding survival rate for hemp is difficult to quantify. This number could range from 90% down to 30%. Factors that will affect seed survival are once again, seeding depth, moisture, temperature. When seeding into favorable conditions, use a higher seed survival number for the seeding calculation. If you are forced into seeding into cool wet or unfavorable conditions, you can lower this number to ensure you obtain an acceptable number of plants ft². This number also takes into consideration the germination rate and seed purity. So, if the germination rate for your seed lot is 90% and you feel conditions for seeding are good, you may use a number of .70 (20% seed mortality). Over 20 yr

of hemp agronomy trials, we've measured hemp seedling mortality rates from a low of 20% to a high of 70%.
Example calculation:

Hemp target plants 10 ft² × 18 grams/1000 kernels ÷.70 ÷ 10.4 = 24.7 lb acre⁻¹

Seeding rates for hemp can and will differ for the following agronomic reasons:

1. Weed control options

2. Conventional or organic production

3. Grain only, dual-purpose, or fiber-only (end use)

4. Row spacing

Table 4 contains general seeding rates and recommended row spacings.

Seeding Date

Overall most hemp varieties require around 110 frost free days to reach full maturity and around 10 inches of precipitation or applied irrigation during this time. Hemp not only dislikes excess water during germination and establishment, but less than ideal soil temperature also has a big effect on hemp plant mortality. During the seeding date trial shown below, three separate varieties

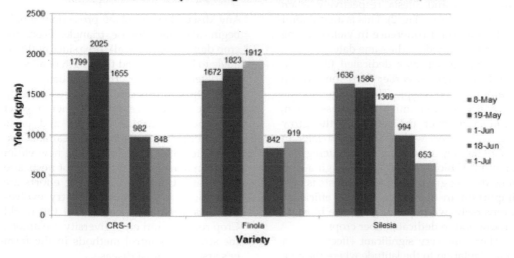

Fig. 2. Hemp grain yields (kg ha⁻¹) as a function of planting date and variety.

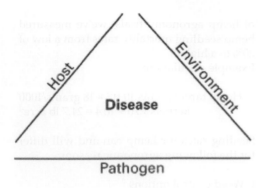

Fig. 3. The disease triangle.

were planted on five separate seeding dates in Manitoba. The varieties chosen were an early maturing grain-only variety Finola, a dual-purpose, moderate-height variety, CRS-1, and a taller dual-purpose tall variety, Silesia. The same seeding depth and seeder was used each time. The seeding density was 23 plants ft². As noted in Fig. 1 above, the early seeding resulted in the highest plant mortality. However, the grain yields were less impacted by seeding date than were mortality ratings. The maximum yields (Fig. 2) were obtained from early to mid-planting dates, whereas plant populations were maximal (mortality minimal) from the 18 Jun planting date.

We note also that there were important responses among the varieties within planting dates. Again, it is imperative to evaluate individual varieties on a regional basis to define potential yields. In this trial, Finola yielded 257 and 543 lbs/A more grain than CRS-1 and Silesia, respectively, when planted on 1 Jun (Fig. 2). This is an economically important difference in yields among varieties planted on the same date.

Hemp grown as a dedicated fiber crop would be planted earlier than some hemp grain or seed crops. This is due to the necessity of maximizing biomass (stem) production over the course of the growing season. As will be discussed later in this chapter, managing or reducing the amount of biomass passing through a combine during grain and seed harvests is very important in increasing harvest efficiency. Conversely, we hope to maximize biomass production in dedicated fiber crops.

There are very significant effects of varietal adaptation to the latitude where the crop is grown relative to the latitude of origin. For example, a variety originally derived from a northern latitude will commence reproductive growth (flowering) much earlier in the growing season when grown at more southern latitudes, thus producing less biomass. Growers of dedicated fiber crops should endeavor to identify and access varieties that are derived from latitudes either closer to or perhaps below the latitude where the crop will be produced (opposite, of course, in the southern hemisphere). These varieties should be most likely to produce maximum biomass.

We must also consider that hemp seed used to produce dedicated fiber crops is derived from production models very similar to hemp grain. But again, we must manage the volume and mass of materials passing through combines during seed harvest. We can manage biomass production from fiber varieties through our selection of planting date. For example, we may plant a dedicated fiber variety early in the season (April or May) to maximize biomass production for fiber. We may plant the same variety much later (June or even July) to significantly reduce the amount of biomass produced (in this case, by reducing the height of the crop), but still achieve an economically viable harvest of hemp seed for propagation of next year's fiber and seed crop. Having a shorter crop relative to a crop planted earlier in the season will significantly improve harvest efficiency of seed from a fiber variety. Harvest protocols for both grain, seed, and fiber crops are discussed later in this chapter.

Disease and Insect Control

Any discussion on plant protection should begin with the disease triangle. Economic crop damage requires all three sides of the triangle to be present and the length of all sides will determine the severity of the damage.

The host in this case would be the hemp crop. Certain cultivars have higher or lower resistance to various diseases and pests. Hemp breeding programs in Canada and the United States are developing new varieties in an attempt to stay ahead of pests and pathogens. Continued breeding efforts are paramount as the hemp industry evolves throughout North America and the world. Crop rotation and crop diversity are among the strongest control methods in the farmer's arsenal against disease.

Environment would include variables such as temperature, soil moisture, and humidity. Besides the season of the year, these factors will be affected by field locations, cropping history, and to some degree, planting density.

In most cases, the pathogen will already be inherently present in most ecosystems. There can be exceptions to this in that new, previously nonexistent pathogens may imported via seed or other propagules (e.g., transplants). The pathogen side of the triangle is control by either chemical, biological, or mechanical methods.

An integrated pest management (IPM) is a means of controlling pests by two or more control methods.

- Cultural control includes practices such as crop rotation, fertilization, seed date, seeding rates, and seeding depth (factors to improve general crop health).
- Biological control would include introducing predators and parasites to control pests
- Chemical control including the use of fungicides, insecticides, and herbicides.

Hemp is a host to several diseases and insect pests. It is important to note that many of these pests do not pose an economic effect on hemp production. In many cases, we still have not determined the economic threshold for damage from these pests.

Currently there are no fungicides or insecticides that are registered for use in hemp crops, so we must rely on cultural and in some cases biological control measures. Data gathered in the past and observations made in the field lean toward using seeding dates and seeding rates to control the major hemp diseases such as, *Sclerotinia, Botrytis,* and *Pythium.*

Sclerotinia sclerotiorum (white mold) and *Botrytis cinerea* (gray mold) are the most common diseases that affect industrial hemp production. Sclerotoia bodies in the hempseed grain are an unsightly contaminant that, like wheat and wild buckwheat, can be difficult to clean out of hempseed and hence will reduce product quality and value.

Sclerotinia or white mold has a very wide range of host crops. It can infect over 400 plant species so therefore can spread quickly from field to field and from plant to plant. White mold needs moist field conditions and the optimal temperature for its growth is 59 °F to 70 °F. Severity of white mold is variable year to year due to its reliance on these conditions for growth, especially high humidity.

Control of sclerotinia is totally dependent on cultural practices as there are no fungicides registered for control in hemp. Reducing environmental conditions that help produce the apothecia through delaying seeding date and lowering seeding rates have all been found to help control sclerotinia.

Proper crop rotation with crops that are not hosts and selection of varieties that are less susceptible are also recommended for control.

The many pathogens that cause damping off prefer many different temperatures, lighting conditions, soil types, and PH ranges. However the common denominator is excess moisture in soils. Damping off can also occur when seed is planted too deep.

Insect Pests

Several insect species have been found feeding on hemp in various growth stages. Field observations have indicated that very few have been found to exceed economic thresholds that would require control. Once again no insecticides are registered for hemp. It is also believed that hemp's natural defense against most insects is credited to the tiny trichromes exuding cannabinoids and terpenes that can act as a natural deterrents to herbivores.

Leaf eaters such as grasshoppers and armyworm have been found in abundance in some hemp crops with little to no effect on hemp yield. Aggressive growth and the abundance

Fig. 4. Sclerotinia disease in a hemp seed head. Photo Credit: Jeff Kostuik

Fig. 5. Mature hemp seed head indicating acceptable harvestability. Photo credit: Jeff Kostuik.

Fig. 6. Hemp along with normal harvest trash in the combine hopper. Photo Credit: Jeff Kostuik

of leaf material assist hemp to withstand damage from high populations of leaf eaters.

European corn borer has been found in hemp fields and can be a concern for farmers that frequently use corn in their crop rotation.

Aphids can often be found in hemp seed heads but once again little is known of their economic damage to grain yield.

Harvest

Regardless of whether growing hemp for grain, fiber, or both, likely THE most challenging undertaking of hemp production is harvesting the crop. From fiber wrapping on combine parts to timing of harvest, hemp harvest is unlike that of any other crop. Simply put, direct experience or a trustworthy friend with direct experience is really the best option to limit issues during harvest activities.

Hemp Grain or Seed Crops

Hemp grain matures roughly in 90 to 120 d depending on many factors including seeding date in relation to latitude, temperature, available moisture throughout the growing season, varieties, and soil fertility–all play a role. In general, hemp plants will flower earlier in a growing season when stressed by environmental (e.g., excess moisture) or edaphic factors (e.g., low fertility). It is not possible to extrapolate the direct effects of stresses on hemp flowering, but it is clear that stressed plants will flower sooner than unstressed plants.

Hemp grain or seed may be harvested as a food for humans or animals (hemp grain), or as seed for the next year's crop (hemp seed). When evaluating maturity of a hemp grain or seed crop, we would evaluate an individual seed head. The seed will become exposed from the bracts that hold it in place as the plant and seed matures. Once approximately 70% of the seeds are exposed in a plant, the seed moisture is likely in the range of 15–25%. Harvest can commence at this time, which is also when about 90% of the seeds will have reached maturity.

The advantage of harvesting seed at this moisture is that the pectin that holds the fibers together in the stalk are still present and are doing their job resulting in less wrapping within the components of the combine. Other advantages of harvesting earlier is limiting the risk of seed shattering due to over maturation and wind. Hemp becomes very prone to seed shatter as it matures and more of the seeds are exposed outside their protective bracts that hold them in place.

The higher the moisture of the grain at harvest, the more plant material will be in your sample which will also contribute to higher grain moisture level. Moisture levels around 12 to 17% are likely ideal, especially for first time growers. This should limit fiber wrapping issues and still give you a relatively clean sample in your grain tank.

Figure 6 is an example of an acceptable sample from a combine hopper. Some leaf material, hurd (stem pieces) and bracts around some seeds will be present. The green material will dry up and not be visible after a few days on aeration in a bin or grain dryer. The sample in this example was testing at 17% moisture.

An obvious disadvantage of harvest at this stage is the need for immediate (within 4 h) airflow through the grain to assist in drying it down. In Canada, bins with aeration fans are used for this purpose and will be discussed further below.

Hemp grain and/or seed is generally harvested with a straight cut header in Canada. It is preferred a draper header is used to feed the hemp stalks in a straight and even manner. Auger headers will work but do not deliver hemp into the combine as well. The cutting knife should be located just below the seed heads to minimize the amount of biomass going through the combine. The cutting knife and guards need to be in excellent shape. Wrapping issues are more prevalent when harvest of the entire stalk is attempted.

Swathing hemp is also an option. The advantage of swathing is that it can provide more control over the grain moisture content. It is recommended to begin picking up the swath within two to three days after cutting so that the stalks do not become too dry, which could lead to increased wrapping of fibers around shafts. Hemp taller than 5 to 5.5 ft tall should not be swathed as this is too much material to run through a harvester.

Either conventional or rotary combines will harvest hemp effectively. Caution should be taken with dual rotor combines because hemp stalks will begin crossing each other inside the combine and will bunch up on the divider separating each rotor.

Individual combine settings will need to be adjusted to crop moisture and harvest conditions. Proper setting of the combine reduces losses and maintains the quality of the hemp grain. Fine tuning the harvest speed and settings is crucial for optimizing your combine.

Suggested initial combine settings:
- Cylinder/Rotor speed: 400–600 RPM
- Concave Clearance: 1–1.2 inch
- Wind: 900–1000 RPM
- Chaffer: 0.4–0.6 inch
- Sieve: 0.1–0.2 inch

Hemp grain is consumed as a raw food product. First and foremost, machinery, trucks, storage bins, and handling equipment must be maintained and free from excess dirt, crop debris, and rodent and bird excrement. Hemp grain should always have strict guidelines for microbial limits including coliforms, mold, yeast. and bacteria. Prevention is much easier than addressing a problem after the fact. It is up to processors to ensure food safety, but food quality and safety begins at the farm level!

Most producers will use grain dryers if the moisture of their hemp crop is above 13 to 14%. Care should be taken whenever handling hemp because the seed coat is sensitive to cracking. The use of conveyors or large augers is recommended. If small augers must be used, slow the auger and keep it full. The temperature of the plenum of the drier should not exceed 140 °F. Hemp grain temperature should never be allowed to exceed 100 °F. Grain toasted in a farm drier is not desirable. Hemp grain is considered dry at 9% moisture. It can then be stored for a period of time safely without further aeration or movement.

Smooth wall bins with hopper bottoms work best for storing hemp that needs to be dried. By easily moving the grain in and out of the bin, one can reduce the risk of hot spots within the bin where grain may not be drying properly.

Aeration fans with adequate horsepower to move air through the grain are used to

A. Maintain grain quality for a period of time

B. Dry grain over a longer period of time

Fig. 7. Standard grain dryers. Photo Credit: Jeff Kostuik.

Fig. 8. Grain storage bins. Photo Credit: Jeff Kostuik.

Fig. 9. Aeration fans used to dry hemp grain. Photo Credit: Jeff Kostuik.

Fig. 10. Supplemental heat source to facilitate rapid drying of hemp grain. Photo Credit: Jeff Kostuik.

Aeration fans can be used to dry hemp grain when moisture levels are less than 13%. Ambient air temperature must be high enough to be effective in drying grain with low humidity levels. Ambient air temperatures are cool, the use of a supplemental heat source will be necessary to assist in drying hemp grain.

As mentioned earlier, cleaning hemp grain to very high standards is essential due to the limited processing that occurs prior to the product going to the end consumer. Cleaning is required for to remove all contaminants such as weed seeds, plant material, insects, and extraneous plant material for both hemp grain and seed. Cleaning should be conducted as soon after harvest as possible. Hemp is marketed as a gluten free crop. Hemp samples should be wheat free to maintain the gluten free status.

Care should be taken in all aspects of handling hemp grain. If the seed coat is cracked, this allows air and oxygen into the seed which causes rancidity and an off taste. Processors will use a peroxide test to determine rancidity of a hemp grain lot.

Hemp Fiber Crops

Hemp crops grown for fiber are generally harvested very soon after female flowering commences. Timing of harvest does have a significant impact on fiber quality (see Chapter 2). As hemp plants enter reproductive growth, additional lignin is produced between fiber bundles. This adds strength to the stem as the flower and eventually the seed are produced, which adds significant weight to the top of the stem. Lignification reduces the probability of lodging, which is the undesirable falling over of mature stems due to the heavy flower/seed head at the top of the stem. Lodging can be a significant problem in all grain crop plants, including hemp.

Increased lignin between fiber bundles is a natural process during reproductive growth. However, more lignin makes separating (also known as opening) fiber bundles significantly more difficult. Additionally, if plants (fibers) are allowed to fully mature, the strength of fibers is reduced relative to less-mature fibers. For these reasons, dedicated fiber crops are harvested when no more than 20% of the crop exhibits female flower development.

The most common method to harvest dedicated hemp fiber crops is with standard hay-making equipment. In general terms, the crop is mowed with a sickle-bar type mower or hay mower which provides for a layer of stems on the ground of a consistent depth (Fig. 11). Stems are then allowed to ret, being turned at least once, preferably with a rotary hay rake.

Once retted, stems are raked into rows and allowed to dry. Once dry, stems are baled with either large square or large round baling equipment and stored indoors or covered to protect from conditions conducive to further retting (moist and warm).

Summary

The agronomic principles used to produce hemp as a modern commodity crop are similar to other commodity crops. However, the key points in successful hemp production are related to a few species-specific characteristics. Variety selection is key to success. Choosing a variety that will provide profitable yields from the desired harvestable component is imperative; that is, a variety bred primarily for grain production may not yield much straw for fiber. It is also important to understand the relationships between the latitude of origin for a variety and the latitude of production. In the northern hemisphere, a variety derived from a northern latitude may begin reproductive growth (flowering) much earlier in the growing season when cultured at a southern latitude. If vegetative growth (high biomass) is the goal as with a dedicated fiber crop, this simple trait could lead to crop failure. Conversely, we can also take advantage of this trait to manipulate plant size at flowering, thus providing for a harvestable crop of seed derived from a variety bred primarily for fiber production. In short, understanding a variety's sensitivity to photoperiod is imperative for profitable yields. We also consider that hemp has extremely poor seedling vigor relative to other commodity crops. This can result in difficulties achieving a profitable stand when environmental or other conditions impose stress during the establishment phase. An example would be untimely precipitation occurring at an excessive rate soon after seeding which could cause soil crusting in a conventional tillage system, or could drown seedlings

(anaerobic conditions) in either conventional/ no-till systems with less well-drained soils. Low seedling vigor, combined with a general lack of available pesticides for hemp crops, define the very strong need for proper field selection for hemp crops (deep, well-drained soils without large, inherent populations of competitive weeds), which will promote successful establishment. Hemp will require added fertility for optimal yields at rates similar to other commodity crops. Once mature, harvesting hemp crops can be extremely challenging, and may require modifications to existing equipment, including post-harvest storage facilities.

Fig. 11. Mowing a dedicated hemp fiber crop with a standard hay mower. Photo credit: Matt Barton, University of Kentucky.

Fig. 12. Raking and baling hemp straw with standard hay equipment. Photo credits: Matt Barton, University of Kentucky.

References

Clarke, R.C., and M.D. Merlin. 2013. Cannabis evolution and ethnobotany. University of California Press, Oakland, CA.

Canadian Fertilizer Institute. 2001. Nutrient uptake and removal by field crops. Canadian Fertilizer Insitute, Ottawa, ON.

Manitoba Agriculture, Food, and Rural Department. 2019. Hemp nutrient utilization. Manitoba Government, Winnipeg, MB. *[2019 is year accessed]*.

University of Kentucky Cooperative Extension. 2018. 2018–2019 Lime and nutrient recommendations. AGR-1. University of Kentucky Cooperative Extension, Lexington, KY.

Chapter 5: Cannabinoids-Human Physiology and Agronomic Principles for Production

R.A. Williams, M.D. and D.W. Williams, Ph.D.*

The plants of the genus *Cannabis* have been drawing human attention for their physiological properties for thousands of years. According to archaeological findings, cannabis has been known in China since the Neolithic period, around 4000 BC (Zuardi, 2006). Shen Nung, a Chinese emperor, is said to be the first to describe the properties and medicinal uses in a compendium on Chinese herbs in 2737 BC (Li, 1974). Use of cannabis spread across India and North Africa, reaching Europe by 500 A.D. (Russo, 2007). Cannabis was brought to the Americas by the Spanish during colonization in 1545 and the English brought it with them to Jamestown in 1611 where it soon became a major industrial interest (Narconon International, 2018). From a western perspective, it was "discovered" in 1839 by a British surgeon working in India with the British East India Company. It was used in British medicine for treatment of appetite disorders, as an anticonvulsant, muscle relaxant, and hypnotic (O'Shaugnessy, 1839; Zuardi, 2006). It was further used by a French psychiatrist, Joseph Moreau, for headaches, to increase appetite, and to aid sleep (Zuardi, 2006). It was listed in 1854 in the U.S. dispensary for medicinal use (Robson, 2001).

The American Marihuana Tax Act of 1937 was enacted to limit use outside of medicine but made obtaining the plant for medicinal or research purposes difficult. This act instituted a tax of one dollar per ounce for medical purposes and $100 per ounce for unapproved purposes (Solomon, 1968). Cannabis was removed from the U.S. pharmacopeia in 1942.

The first cannabinoid was isolated as cannabinol in 1898, and further research elucidated more cannabinoids in the 1960s (LaPoint, 2009). There was significant pharmacologic interest in the compounds through various research projects in the 1960s, although it proved difficult to eliminate the unwanted psychoactive side effects of the compounds. Further research and elucidation of the CB 1 and 2 receptors as well as discovery of synthetic and endogenous cannabinoids led to more targeted formulations and utilizations of the cannabinoids. This research continues today.

Obviously, with pharmaceutical use and known intoxicating effects, there has been historical abuse of the cannabis plant as well. This has been well publicized in mass media, has been the subject of a significant amount of legislature both in the United States as well as abroad, and has garnered a significant amount of law enforcement attention. The Boggs Act and Narcotics Control Act of 1951

R.A. Williams, University of Kentucky, College of Medicine; D.W. Williams, University of Kentucky College of Agriculture. *Corresponding author (dwilliam@uky.edu)

doi:10.2134/industrialhemp.c5
© ASA, CSSA, and SSSA, 5585 Guilford Road, Madison, WI 53711, USA.
Industrial Hemp as a Modern Commodity Crop. D.W. Williams, editor.

increased penalties for cannabis possession and distribution to include mandatory prison sentences (Bonnie and Whitebread, 1970). The Controlled Substances Act in 1970 listed marijuana, along with LSD and heroin, as a schedule I drug, a drug with the highest abuse potential without potential medical use (Zeese, 1999). This has led to significant controversy and greatly increased difficulty with research regarding possible scientific uses of the compounds contained in cannabis. "Zero tolerance" policies by the Reagan and Bush administrations led to the Anti-Drug Abuse Act of 1986, which reinstated mandatory sentences for possession alone and increased governmental efforts to prevent smuggling of cannabis across U.S. borders (Lee, 2012). In 1996, California became the first state to legalize smoked and edible cannabis for use by people with AIDS, various cancers, and other serious illnesses (Lee, 2012). Several states followed suit. In 2012, Colorado Amendment 64 and Washington's Initiative 502 made these respective states the first to legalize cannabis for recreational use. Since then, a majority of states have legalized medical uses of cannabis, and several others have also legalized recreational uses.

Internationally, Uruguay was the first country to fully legalize cannabis on a nationwide level. Their laws permit acquisition and use by growing a limited number of plants, purchasing governmentally produced products from pharmacies, or joining cannabis clubs that are permitted limited growing capacity. Canada legalized all uses of cannabis in 2018. In the Netherlands, cannabis has been available since 2003 in standard concentrations and dispensed in pharmacies (Gorter et al., 2005). Cannabis use remains illegal in most European Union countries, with penalties for violations varying widely. Generally, relaxation of restrictive legislation as well as increases in societal acceptance of the use of cannabinoids both as a pharmaceutical compounds as well as a recreational intoxicants has led to an increased ability to study the plant for its wide spectrum of pharmaceutical and industrial uses.

There are 460 chemicals identified in the two major species of *Cannabis*, *Cannabis sativa* L. and *Cannabis indica* L. Of these 460 chemicals there are multiple types of bioactive molecules including flavonoids, terpenes, alkaloids, and cannabinoids. "Cannabinoids" is the name given to over 80 oxygen containing C-21 aromatic hydrocarbon compounds (ElSohly et al., 2005) that have so far been isolated from cannabis. The term has also been extended to include similar chemical molecules which are synthetically derived as well as including various other molecules which act at cannabinoid receptors as opposed to a rigid, chemical structure-based definition. As such, the term phytocannabinoids has been utilized to describe the organically, plant-synthesized molecules to avoid confusion (Pertwee, 2005).

Δ-9-tetrahydrocannabinol (Δ-9-THC) is the major intoxicating constituent of cannabis. Other commonly referenced constituents include: Δ-8-tetrahydrocannabinol (Δ-8-THC), cannabinol (CBN), cannabidiol (CBD), cannabicyclol (CBL), canniabichromene (CBC), cannabigerol (CBG). Please note the abbreviations for these constituents as they will be used frequently in this chapter. As far as is known today, the constituents outside of Delta-9-THC have very little intoxicating effect and are generally present in smaller concentrations (Smith, 1998) but are thought to contribute to the overall effect of cannabis (Ashton, 2001).

Biosynthesis

Cannabinoids can be isolated from all tissues of the cannabis plant but are mainly produced in the trichomes of the leaves and flowers, but mostly the flowers of the female plant (Turner et al., 1978). Biochemical synthesis was studied initially in the 1970s using radiolabeling to further elucidate the pathway (Shoyama et al., 1975). There are three basic steps to the biochemical pathway: binding, prenylation, and cyclization. The binding step includes simple protein binding to initiate the combination of the initial phenol component, olivetolic acid (OA), and the terpene component, geranyl diphosphate (GPP). The latter of these components, GPP, is obtained from the combination of dimethylallyl diphosphate (DMAPP) and isopentenyl diphosphate (IPP) via the DOXP pathway utilizing the enzyme GPP synthase (Fellermeier et al., 2001). Olivetolic acid is obtained from the combination of hexanoyl-CoA and hexanoic acid obtained by multiple levels of hydrolysis via the polyketide pathway (Di Marzo et al., 1995). Geranyl diphosphate and OA are then combined with the help of an enzyme named cannabigerolic acid synthase (CBGAS) (Morimoto et al., 1998), which is a GOT family enzyme, causing the alkylation of OA by GPP

(Fellermeier and Zenk, 1998). The resulting cannabigerolic acid is then subsequently synthesized into the other cannabinoids.

With heat decarboxylation only, the CBGA loses a carbon dioxide molecule and is therefore converted into cannabigerol, the active cannabinoid. More aromatic rings can also be added to the one in CBG (from olivetolic acid) by a group of enzymes called oxidoreductases causing electron transfer from the chemical donor to the electron acceptor. This is termed cyclization and creates THCA, CBDA, or CBCA depending on the presence and relative quantities of the specific, requisite enzymes (Gaoni, 1964; Fellermeier, 2011). The presence or absence of these enzymes is what leads to the differing quantities of the specific cannabinoids within different chemptypes of plants.

Cannabigerolic acid is converted to Δ-9THCA by THCA synthase (Taura et al., 1995) by cyclizing two aromatic rings from olivetolic acid. The presence of this enzyme is the factor that drives the difference between high THC containing cannabis, or "marijuana", from minimally intoxicating cannabis, or "hemp". Plants which are high in THC concentration have a different form of the active gene encoding for the THCA synthase enzyme than the generally inactive THCAS gene in non-intoxicating plants (Kojoma et al., 2006). Cannabigerolic acid synthase catalyzes a reaction where a new carbon-to-carbon bond in a similar reaction to that which forms THCA, however, now missing is the new carbon-to-oxygen bond, thus creating CBDA instead of THCA (Fellermeier et al., 2001). Cannabigerolic acid is formed by causing a carbon-oxygen bond at a separate position than is utilized in the THC or CBD reaction causing a new bicyclic molecule. This is catalyzed by CBCA synthase (Fellermeier et al., 2001). As a result, cannabinoids are mostly found in the acidic component in the dried plant, but then undergo the same heat decarboxylation either when cooked or smoked to form their active constituents (Baker et al., 1981; Bosy and Cole, 2000). Delta-8-THC is formed by the degradation, or more specifically, the isomerization of Δ-9 THC changing the carbon–carbon double bond position in the THC ring (Holler et al., 2008). This is not an enzymatic driven reaction, but occurs spontaneously. Furthermore, both Δ-8-THC as well as Δ-9-THC can be degraded into cannabinol (CBN). This occurs by removing four hydrogens from the first carbon ring of the molecule to create an aromatic ring (McCallum et al., 1975). The diversity of the remaining cannabinoids compounds is a result of nonenzymatic reactions.

Bioactivity

The bioactivity of these molecules in human beings has been under study for several years. Initially, the cannabinoids were thought to perturb cell membranes, but later they were found to act on specific receptors at the cell membrane (Devane et al., 1988). The two main cannabinoids receptors have been termed CB 1 (Devane et al., 1988) and CB 2 (Munro et al., 1993). Both of these are G-coupled receptors, inhibiting adenylate cyclase and protein kinase. They are stereospecific and dose-dependent (Pertwee, 2005; Howlett et al., 2002). There has also been evidence that these receptors may affect calcium and potassium flux as well (Pertwee, 2005). CB 1 receptors have been found heterogeneously in human tissues but are most highly concentrated in both the central and peripheral nervous systems (Pertwee, 2005; Howlett et al., 2002). Activation of the CB 1 receptor in humans causes cognitive impairment, dyskinesia, and analgesia (Pertwee 2005). CB 1 receptors have also been appreciated at the nerve terminals of the neuron both centrally and peripherally, causing modulation of neurotransmitters at this site (Howlett et al., 2002, Pertwee, 2005). Given the neuronal involvement and widely-espoused psychoactive effects of THC, the CB 1 receptor is more widely studied (Martin and Wiley, 2004). Less is known about the CB 2 receptor. CB 2 receptors have been found mostly in the immune cells (Howlett et al., 2002; Pertwee, 2005). CB 2 activation seems to modulate immune function, regulate cytokine release, and affect migration of activated immune cells (Pertwee, 2005). Non-CBD receptors can also be activated by cannabinoids with many different effects including neuronal inflammatory signaling, vascular contractility, gastrointestinal muscle tone, immune cell migration, cytokine release, antioxidant effects, as well as anti-inflammatory and antitumor properties (Pertwee, 2005).

The agonists for CB receptors have been separated into four main categories: classical, nonclassical, aminoalkylindole (synthetic), and eicosanoid (Howlett et al., 2002). The classical denotation is given to the phytocannabinoids or their derivatives including THC, CBD, and CBN. Tetrahydrocannabinol is highly lipid soluble and is a partial agonist for both receptors with lower affinity than some other cannabinoids. Cannabinol is a partial agonist at CB 1, but has less activity than THC (Pertwee, 2005). Cannabinol may, however, bind and activate CB 2 more effectively than THC (Pertwee, 2005). Cannabidiol is thought to have very low affinity for both CB 1 as well as CB 2 receptors (Pertwee, 2008) but is thought to have effects outside of CB receptor activation (Howlett et al., 2002). These effects are thought to include vascular dilation, microglial cell migration (Pertwee, 2005), anti-oxidant activity, and regarding the release of many different neurotransmitters (Pertwee 2005).

The nonclassical category includes synthetic compounds very similar in structure to the phytocannabinoids but lacking a pyran ring (Melvin et al., 1984). They were initially developed by Pfizer in the 1980s as possible analgesics. These molecules are designated with the nomenclature CP which is an abbreviation for Charles Pfizer himself. This class of compounds were among the first synthetic cannabinoids to be identified in formulations which were marketed as "herbal incense". They have high affinity for both CB 1 and CB 2 receptors (Howlett et al., 2002). They are no longer used in any pharmacological preparations as the U.S. Drug Enforcement Administration (DEA) used emergency scheduling to control these compounds in March 2011 after widespread abuse. They are found to be 10 to 50 times more potent in activation than the phytocannabinoids when studied in mouse models (Johnson and Melvin, 1986).

Aminoalkylindole agonists are starkly different in structure than the other cannabinoids (Bell et al., 1991). They have high affinity for both CB 1 and CB 2 receptors (Bouaboula et al., 1997). This is the most common class of compound abused in synthetic marijuana. These are herbal products that are laced with synthetic cannabinoids. Notable members of this classification include aphthoylindoles (e.g., JWH-018), phenylacetylindoles (e.g., JWH-250), and benzoylindoles (e.g., AM-2233 and UNODC, 2011). JWH-018 is the most widely known synthetic compound of abuse and is considered to be three times more

potent than phytocannabinoids. It was developed as a test compound in the lab of John William Huffman during his research on cannabinoid receptors (Wiley, 2011).

The eicosanoid class includes endocannabinoids, which is the term given to mammalian-derived compounds of similar receptor action to the phytocannabinoids, but have markedly different structures from the other classes previously discussed (Howlett et al., 2002; Pertwee, 2005). These compounds are produced in mammalian tissues and are mostly free fatty acid derivatives of arachidonic acid (Di Marzo et al., 1995). They act in a retrograde fashion as messengers at the neuron synapse. Neurons are believed to react to GABA release and subsequent increase of cAMP to synthesize endocannabinoids on demand (Pertwee, 2005). They are believed to enter the cell by both diffusion and carrier-mediated transport subsequently decreasing adenylyl cyclase, decreasing cAMP, and therefore decreasing calcium influx and potassium efflux from the cell. The net result of this action is hyper polarization of the neuron and subsequent decreased synapse firing (Pertwee, 2005). The most investigated of these compounds include anandamide and 2-arachidonoyl glycerol. Anandamide acts as a partial agonist for both the CB 1 and CB 2 receptors (Pertwee, 2005), with greater CB 1 affinity as compared to CB 2 (Howlett et al., 2002). 2-arachidonoyl glycerol activates both receptors, binds equally well, and seems to have higher efficacy when compared to anandamide (Pertwee, 2005).

Cannabinoids are lipid soluble and seem to show a steady state distribution (Grotenherman, 2003). The compounds are distributed into a highly vascularized tissues first and are stored in adipose tissue (Nahas, 1971). Tetrahydrocannabinol has been shown to cross the placenta when administered to pregnant rhesus monkeys (Bailey et al., 1987). Likewise, THC has also been found in breast milk of lactating mothers (Perez-Reyes and Wall, 1982).

Regarding metabolism, THC undergoes hydroxylation and oxidation in the liver, controlled by CYP enzymes (Watanabe et al., 2007). The primary metabolite is 11-OH-THC, which is active, and further oxidized into the multiple inactive metabolites (Agurell et al., 1986). THC has a wide-ranging half-life ranging anywhere from 1.6 to 57 h (Grotenherman, 2003). Metabolites are then excreted 15% of the urine and 50% in the feces (Busto et al., 1989). Measured at 5 d, 80 to 90% of the concentration of THC is excreted (Hawks 1982). In chronic users, metabolites can be detected in the urine for several weeks, modified by age, weight, and frequency of use (Ellis et al., 1985).

As noted above, the physiological effects of cannabinoids are wide-ranging. The psychological effects include relaxation, perceptual alteration of both sensory and temporal components, euphoria, as well as increased appetite (Grigoryev et al., 2011). Corporal effects include increased cerebral perfusion (Mathew et al., 1997), increased heart rate as well as decreased vascular resistance (Jones, 2002), decreased airway resistance (Tashkin, 2001), and decreased intraocular pressure (Mikawa et al., 1997). Adverse effects may include lethargy, hypoplasia, discrimination, and impaired cognition. Cannabinoids can cause feelings of paranoia, dysphoria, xerostomia, dizziness, sedation, postural hypotension, abdominal discomfort, as well as nausea and vomiting. Cannabinoids have been found to be cardioactive and there have been case reports of dangerous cardiac dysrhythmias, but these studies are confounded secondary to patient history of previous cardiac disease (Rezkalla et al., 2003). Acute toxicity of inhaled cannabis is usually very mild, but can be life-threatening in children causing tachycardia, apnea, cyanosis, bradycardia, hypotonia as well as opisthotonos (Macnab et al., 1989). Synthetic cannabinoid toxicity is more difficult to ascertain as these formulations, as abused, are usually tainted with other xenobiotics, mostly stimulants.

Chronic cannabinoid use has also shown toxicity. Chronic obstructive pulmonary disease (COPD), as seen with tobacco use, has been shown in people participating in the inhalation of cannabis (Wu et al., 1988). There may be an increased risk of coronary artery disease as well as myocardial infarction (Mittleman, 2001). A vomiting syndrome, termed cannabinoid hyperemesis syndrome, has been elucidated, although the mechanism is unknown. This causes profuse and unrelenting vomiting and abdominal pain, which may be refractory to classic treatment such as opiate pain medications as well as classic antiemetics (Wallace et al., 2011; Galli et al., 2011). A myriad of psychiatric issues have been proposed as chronic side effects of chronic cannabis use although little data has been provided as evidence.

Utilization in Modern Medicine

Cannabis-derived substances have garnered much interest for treatment of a spectrum of pathologies in medicine, and use has become more popular and widely accepted by physicians. A highly-debated 2005 study polled physicians indicating about 36% supported legalization and 26% were neutral (Charuvastra et al., 2005). A survey in 2013 showed 76% of physicians polled approved marijuana for medical purposes with most physicians stating "the responsibility as caregivers to alleviate suffering" as their reason (Adler and Colbert, 2013). The American Medical Association has stated that it would support rescheduling if it would facilitate research and development of cannabinoid-based medicine (Hoffmann and Weber, 2010). Clinical studies have been especially difficult secondary to blinding the psychoactive effects of cannabis as a whole. There is a significant amount of conflicting data and bias with research being backed by financially-interested parties. Despite these difficulties, there continues to be a significant amount of research regarding numerous potential therapeutic mechanisms.

Data in the use of cannabinoids for pain control has been conflicted. Cannabinoids have been shown to have synergistic effects with opioids in the alleviation of pain (Abrams, 2006), but older studies have shown no benefit of cannabinoids in the treatment of acute pain (Jain et al., 1981). An oral spray with equal parts CBD and THC has shown benefit over placebo in a randomized controlled trial which included 360 patients (Portenoy et al., 2012). A similar combination of compounds was investigated in a randomized controlled trial with 177 patients showing benefit over placebo although a formulation with THC alone showed statistically equal efficacy but no significant improvement over placebo (Johnson et al., 2010). Cannabinoids have been specifically studied in the palliation of neuropathic pain. It is believed to work on neuron C-fibers to modify the pain response in hyperalgesia (Manzanares et al., 2006).

There are a multitude of reports that show preclinical support and efficacy, but much of this data is of questionable quality, secondary to the fact that many patients are utilizing herbal extracts as opposed to directed chemical agonists. Although a 2015 meta-analysis in the Journal of American Medical Association cited moderate-quality data for chronic pain (Whiting et al., 2015), further studies in the use of THC alone have shown no efficacy in the treatment of chronic non-cancer pain with the number needed to benefit 24 and the number needed to harm 8 (Stockings et al., 2018).

Along with analgesia, the antiemetic properties of cannabis are probably the most widely used and popularized. Traditionally, corticosteroids, serotonin receptor antagonists, neurokinin receptor antagonists, and antipsychotics have been utilized to help palliate the profound nausea and vomiting that frequently is a side effect from both oral and parenteral chemotherapy. These drug classes have varying efficacy, and each has its own respective significant side effects. As such, cannabinoids have been investigated both as an adjunct therapy as well as use in isolation.

Cannabinoids are thought to act through acetylcholine activity in the brain (Coutts and Izzo, 2004), specifically at the nucleus tractus solitarii at the level of the area postrema—the area of the brain believed to be active during the emetic reaction (Himmi et al., 1996). The two most utilized and studied formulations are Nabilone and Dronabinol (synthetic forms of THC). Unfortunately, the subsequent efficacy data here is again conflicted. Cochrane review data has shown positive nausea and vomiting response with moderate quality data (Whiting et al., 2015). A systematic review of the use of synthetic cannabinoids has shown increased emetic control when compared to older antiemetics including prochlorperazine, metoclopramide, chlorpromazine, and haloperidol with a low number needed to treat for full emetic control (Tramèr et al., 2001). Meta-analysis of over 600 patients utilizing nabilone and dronabinol shown increased efficacy for emetic control over these same routinely used antiemetics (Ben Amar, 2006). When utilized as adjunctive therapy with highly emetogenic chemotherapy such as high dose methotrexate, cisplatin, doxyrubicin, and cyclophosphamide, the cannabinoids were not shown to be more effective than traditional therapy. They caused a significant amount of adverse sequelae including intoxication (30%) as well as sedation (20%).

Interestingly, during the study, the physicians allowed for treatment arm crossover and found that patients preferred cannabinoid therapy for future treatment cycles in a 90% majority (Tramèr et al., 2001). It seems that the use of cannabinoids is generally viewed as less attractive than the newer serotonin and neurokinin receptor antagonist therapy which are regarded as more potent, having no psychoactive effects, and can be administered intravenously requiring no ability to tolerate oral intake (Davis, 2016). A randomized controlled trial by Plasse et al. (1991) as well as Lane et al. (1991) showed dronabinol increased control and nausea and vomiting when combined with prochlorperazine; however, a similarly-performed randomized controlled trial by Meiri et al. (2007) showed no such significance when dronabinol was compared with ondansetron (a very commonly prescribed serotonin antagonist antiemetic). Currently, although cannabinoid therapy has shown some promise, the trend is for utilization as adjunct therapy in refractory cases.

Similar to chronic pain and nausea and/or vomiting, malignancy and subsequent treatment with chemotherapy has shown to have significant anorexia as another noteworthy physiologic side effect. Similar anorexia can be appreciated in other serious illnesses including HIV and AIDS. Herbal extracts of cannabis, cannabis ingested as a whole plant therapy, as well as various pharmacologic formulations have been utilized and studied in the attempted palliation of this side effect. Unfortunately, the results of these investigations of also been mixed. A Cochrane review of 7 randomized controlled trials in patients with HIV/AIDS found variable efficacy regarding appetite, weight, performance, and mood (Lutge et al., 2013). Randomized controlled trial of 243 patients with anorexia and subsequent cachexia in the context of malignancy showed no superiority of either cannabis extract or THC over placebo (Strasser et al., 2006). A double blinded randomized controlled trial of a larger population of 469 patients found that a popular non-cannabinoid medication, megestrol acetate, utilized in the treatment of anorexia showed more efficacy than dronabinol in stimulating appetite and increasing weight gain. Furthermore, as adjunctive therapy with THC was shown to be no more effective than megestrol acetate alone (Jatoi et al., 2002). Still, dronabinol is frequently utilized in the stimulation of appetite caused by various pathologies and continues to be championed anecdotally.

In addition to symptom control, there has been significant research in the use of cannabinoids as a cytotoxic agent in the treatment of cancer. Many cancers have displayed cannabinoid receptors which have been related directly to the degree of dysplasia, and therefore, the severity of tumor (Davis, 2016). Cannabidiol has shown cytotoxic activity in breast cancer (Ligresti et al., 2006), various skin cancers (Baek et al., 1998), and has mechanistically been shown to act on a target that has been related to prostate cancer (De Petrocellis and Di Marzo, 2010). Interestingly, cannabinoid antagonists have also been shown to have antitumor properties (Davis, 2016) further showing evidence that the cannabinoid system is involved with cancer metabolism.

Outside of malignancy related therapies, the antiepileptic property of various cannabis preparations is arguably most promising. Older randomized controlled trials have shown efficacy in the control of seizures (Gloss and Vickrey, 2014). Anecdotally, a survey of parents belonging to a social media group which focused on the use of CBD-enriched cannabis for control of seizures in children with early-onset severe forms of epilepsy found that 16/19 parent respondents reported reduced seizure frequency in their children. Two reported complete seizure freedom and eight reported a greater than 80% reduction in seizure frequency (Porter and Jacobson, 2013). On 25 June 2018, the FDA granted approval of Epidiolex (98% CBD oil-based oral solution) in patients two years and older for the treatment of seizures resulting from Dravet and Lennox-Gastaut syndromes. Seizures have been notoriously difficult to control leading to significant morbidity and mortality in these pathologies (Devinsky, 2016). Double-blind randomized controlled trial with 120 pediatric patients with Dravet syndrome showed significant efficacy of Epidiolex causing decrease of seizures by almost half and causing 5% of children to become seizure free (Devinsky et al., 2017). The study, as well as significant anecdotal support, has led to the FDA granting priority review, fast-track, and orphan drug designation, which is a rare occurrence. Epidiolex is currently under investigation for other treatment resistant forms of epilepsy such as Febrile Infection-related Epilepsy Syndrome (Gofshteyn et al., 2016). Side effects are similar to other forms of cannabinoids- diarrhea, vomiting, fatigue, pyrexia, somnolence, and abnormal results on liver-function tests (Devinsky et al., 2017).

Concurrent to approval by the FDA, the DEA rescheduled Epidolex from a schedule 1 to a schedule 5 controlled substance. Rescheduling allows for physicians to prescribe appropriately, but it also allows for the use of Epidiolex in clinical studies under far less onerous regulations than with schedule 1 compounds. Undoubtedly, this will drastically increase clinical research with CBD in the form of Epidiolex for many potential pharmaceutical uses; a very exciting outcome for those most interested in medical research with cannabinoids.

There is ongoing research regarding the benefit of utilizing cannabinoids in the treatment of traumatic brain injury. This interest has increased given the recently-increased attention to repeated head injuries both in traumatic sports as well as the significant increase in traumatic head injury data from recent military operations around the world. There is some data showing the endocannabinoid system plays a role in regulating inflammation after traumatic brain injury with CB-2 agonists reducing inflammation and CB-2 antagonism accomplishing the opposite (Amenta et al., 2014). Studies in animal models have shown cognitive benefit after traumatic brain injury (Reiner et al., 2014) and rat model data shows that cannabinoids (both THC and CBD) are potentially neuroprotective (Hampson et al., 1998). Some retrospective and anecdotal data when evaluating traumatic brain injury patients which have used THC recreationally prior to their injury has shown improved neurologic outcome (Nguyen et al., 2014). The current data is promising but requires further elucidation and continued investigation.

There have been multiple studies regarding the use of cannabinoids in the treatment of dystonic and spastic disorders such as multiple sclerosis, Huntington's disease, Alzheimer's dementia, Parkinson's disease, and others. A systematic review identified 14 placebo-controlled trials with over 2000 participants with spasticity (Whiting et al., 2015). The studies included patients suffering from symptoms of multiple sclerosis, as well as paraplegia caused by spinal cord injury. They investigated Nabiximols (an herbal extract of cannabis utilized as a pharmaceutical internationally), dronabinol, nabilone, THC/CBD, ECP002A (THC) and smoked cannabis. Another systematic review by the American Academy of neurology included 17 studies consisting of over 1000 patients. The analysis showed efficacy in patients' self-reported spasticity. However, there was no effect on the objective resistance to passive stretching of the soft tissues (Koppel et al., 2014). The six studies that included information on patient tremor found no improvement with THC, oral cannabis extract, or nabiximols. Smaller studies regarding Parkinson's disease have shown some symptom amelioration (Lotan et al., 2014; Sieradzan et al., 2001) and one small randomized controlled trial showed improvement of Parkinson's disease symptoms (Carroll et al., 2004). There are currently no clinical trials underway for Alzheimer's dementia, but there is some mechanistic evidence showing neuroprotection and possible myeloid plaque reduction, the underlying physiologic mechanism for the disease (Ahmed et al., 2015; Martin-Moreno et al., 2011). Overall, evidence in use of spastic disorders has been mixed and clinical utility is still suspect.

Glaucoma, an increase in the intraocular pressure of the eye causing potential blindness, has been another disease which cannabinoid therapy has been widely popularized. A 1971 study performed by Hepler and Frank showed a decrease in intraocular pressure by 25 to 30% but they noted that the effects are short-lived and require frequent dosing. Both smoked cannabis as well as synthetic cannabinoid formulations both lower intraocular pressure, with the synthetic formulation not showing significant adverse effects (Green, 1998). Ocular pathologies are particularly appropriate for topical administration, allowing for minimal systemic effects. Unfortunately, cannabinoids have shown to have poor ocular penetrance (Järvinen et al., 2002). There has been some success in improving penetrance by adding cyclodextrins and formulating various emulsions (Naveh et al., 2000), but cannabinoid therapy is generally viewed as inferior to current traditional therapies secondary to both the wide-ranging adverse systemic effects and non-superiority of cannabinoids.

The well-known psychoactive effects of cannabinoids have led to increased interest in the potential treatment of various psychiatric pathologies. A randomized controlled trial of 10 patients with generalized social anxiety disorder found that 400 mg of CBD significantly decrease their anxiety (Crippa et al., 2011). Two randomized controlled trials with generalized social anxiety disorder found that pretreatment with CBD reduced anxiety associated with a simulated public speaking test (Bergamaschi et al., 2011; Zuardi et al., 1993). Retrospective chart review of 80 posttraumatic stress disorder patients showed that cannabis use shows reduce symptom scores when utilizing an objective patient reported symptom scale (Greer et al., 2014), with an additional retrospective chart review showing 72% of PTSD patients using nabilone had reduced or eliminated nightmares (Fraser, 2009). A randomized treatment-controlled crossover trial in fibromyalgia patients with chronic insomnia found that nabilone before bedtime was superior to amitriptyline, a commonly prescribed antidepressant used for sleep and for improving sleep quality (Ware et al., 2010). A placebo-controlled study in obstructive sleep apnea patients also found that dronabinol was superior to placebo regarding objective sleep apnea measures (Prasad et al., 2013).

Serotonin metabolism in the nervous system has been widely implicated in the pathology of anxiety and depression. Cannabinoids have been shown to interact with serotonin signaling (Russo, 2011), and as such have been investigated in the treatment of these pathologies. In rodent models, CBD has been shown to ameliorate subjective signs of anxiety (Russo, 2011), decrease subjective fear behaviors (Resstel et al., 2009), decrease subjective depressive-like symptoms in PTSD (Burstein et al., 2018), and generally show anti-depressant effects (Zanelati, 2010). Cannabidiol may also have antipsychotic effects by CB-1 antagonism, and is thought to potentially help ameliorate addiction to marijuana and tobacco (Mandolini et al., 2018). Unfortunately, the studies investigating psychiatric therapeutic potential are limited secondary to small sample size as well as difficulty in objectively measuring study outcomes. More data will be needed before cannabinoid-based therapy becomes widely accepted.

Cannabinoids have been investigated regarding a myriad of other physiological effects as well. Regarding metabolic effects, phytocannabinoids are reported to produce decrease body fat, decreased leptin (a hormone which regulates hunger) and increased energy expenditure in a mice model (Riedel et al., 2009). Multiple similar studies have stimulated interest in possible pharmacologic targets for diet and weight control. Cannabinoids have been shown to have antibiotic effects with activity against multi-drug-resistant *Staphylococcus aureus* (Appendino et al., 2008). Cannabigerol specifically has shown antibacterial as well as antifungal effects (Elsohly et al., 1982). Cannabinoids have been shown to have a bronchodilatory effect on respiratory physiology (Williams et al., 1976). Tetrahydrocannabinol has been shown to have anti-inflammatory action which is 20 times that of aspirin, a commonly used non-steroidal anti-inflammatory, and twice that of hydrocortisone, a commonly used corticosteroid, when studied in standard animal tests and cellular assays (Evans, 1991). Further inflammatory modulation has been observed through interaction with T cells and cytokines (Nagarkatti et al., 2009). Further investigation into these less popularized therapeutic mechanisms will be necessary before significant applications are ready for clinical adoption.

Cannabinoids do not seem to act without synergy within the class and in addition to other compounds. Synergism between compounds in the cannabinoid class was first theorized when smoked marijuana was found to have more significant effects than isolated THC alone (Carlini et al., 1974). This is termed "entourage effect". This has been further evidenced by newer studies showing that CBD alone requires narrower therapeutic windows and higher doses than a whole plant preparations, implying synergistic actions between the various phytocompounds (Gallily et al., 2015). Cannabiodiol has been shown to inhibit the human liver enzyme, CYP2c9, which metabolizes THC, causing subsequent increase in serum THC levels (Yamaori et al., 2012). CBD can also act as an antagonist and allosteric modifier to decrease the activation of THC. CBD pretreatment has been shown to raise the cerebral level of THC by threefold (Reid and Bornheim, 2001). A corporation primarily involved in the research of cannabis recently released a press statement that they have isolated three additional cannabinoids that directly increase the effects of THC (BusinessWire, 2016). Symptomatically, concurrent CBD use has been shown to reduce anxiety produced by THC alone (Williamson and Evans, 2000).

Cannabinoids seem to also interact with other terpene compounds. Cannabinoid interaction with myrcene (a mono-terpinoid) was shown to cause greater sedation when compared with the sedative effect by either compound alone (Russo 2011). Cannabinoids also seem to interact with various commonly-studied terpenes including limonene, linalool, and pinene, causing a myriad of physiological effects (Russo 2011). Less common phytocannabinoids have also been shown to interact with terpenes to work in synergism, causing a multitude of effects including: antibacterial, antifungal, anticonvulsant, dermatologic, analgesic, and immune modulation (Russo 2011).

Cannabinoid interaction with non-terpene compounds has also been studied. As previously discussed, cannabinoids are metabolized in humans by the liver utilizing the CYP enzyme system. The metabolism of cannabinoids could secondarily affect the metabolism of other compounds also metabolized in a similar pathway (Grotenhermen, 2005). Furthermore, THC has been shown to be a CYP1A2 inhibitor, potentially decreasing serum concentrations of other commonly-used adjunct medications across a variety of drug classes (Yamaori et al., 2010). CBD inhibits the liver enzyme CYP 3a4 and CYP 2d6. CYP 3a4 metabolizes about a quarter of all medications, and as such concurrent use of CBD would potentially change serum concentrations of popularly-prescribed medications including macrolide antibiotics, hypertension medications, benzodiazepines, antihistamines, antiretrovirals, and some cholesterol medications (Watanabe et al., 2007). This could lead to effects regarding efficacy and potential side effects of these medications, as the CYP system is utilized not only in the elimination of medications but also metabolism of nonbioactive prodrugs to the active form of some medications.

Cannabinoids have been shows to have additive effects when administered concurrently with other pharmaceuticals. Increased CNS depression, drowsiness, and somnolence have been observed with concurrent use of cannabinoids and other CNS depressants (Kumar et al., 2001). Additive analgesia has been described when cannabinoids have been used concurrently with opiate pain medications, independent of effect on opiate serum concentration and without increased sedation (Abrams and Guzman, 2015). Tetrahydrocannabinol and CBD have been shown to increase warfarin (a commonly utilized blood thinner), thereby altering patient coagulation (Yamaori et al., 2012). Cannabinoids have been purposed to interact with the metabolism of several anti-psychotic medications, increasing clearance (thereby decreasing efficacy) of chlorprom-azine (Wills, 2005) and increasing serum concentrations (thereby increasing efficacy) of others (Zullino, 2002). Concurrent use may also alter serum levels of some impor-tant anti-viral medications (Wills, 2005).

Cannabinoids have been shown to inter-act with other substances of potential abuse as well. Alcohol, when used concurrently with cannabinoids, increases THC levels and has shown increased impairment of both cognitive and motor function while driving (Hartman et al., 2015). Cannabidiol increases the uptake as well as the efficacy of some potential drugs of abuse such as PCP and cocaine but does not seem affect phar-macokinetics of others such as morphine and MDMA (Reid and Bornheim, 2001).

Summary of Human Physiology

Despite conflicted clinical evidence, we will likely see continued interest in investigation of cannabinoids as there is a positive trend for legalization of cannabis sparked by financial gains by increased tax revenue at the gov-ernmental level as well as cost savings from decreased legal actions and enforcement of marijuana laws. In the future, further inves-tigation in the dosage forms, administrative routes, drug–drug interactions, appropriate dosing, and safety concerns will be neces-sary (New Frontier, 2017). There will likely be investigation into non-CBD receptor pharma-cologic targets. Alternate receptor sites have been shown to change cell migration targets in disorders such as endometriosis (McHugh et al., 2010). Furthermore, elucidation in the cross-reactivity between cannabinoid recep-tor targets such as interactions between the cannabinoid agonists themselves as well as cannabinoid allosteric activity in the pathways of other chemical receptors such as dopamine and the other neurotransmitters may prove to be quite useful. Some current investigations

in restricting CBD targets to peripheral acting receptors have shown some therapeutic ben-efits with little side effects (Zhang et al., 2018). As these treatments become more widely accepted, continued randomized controlled trial studies will be necessary to further elu-cidate and describe the exact efficacy and side effects in the use of the various routes and formulations of cannabinoids versus placebo and traditional treatments. Further-more, further research on the production of these compounds will be necessary to make a financially viable supply for both research and consumption. This could include production via cell culture or perhaps transgenic plants.

In conclusion, the human physiology of cannabinoids is wide ranging both toxico-logically as well as a pharmaceutically. The molecules work in a myriad of pathways and on multiple receptors, causing a multitude of very diverse effects. Cannabinoids have been studied in many pathologies, but only a few have gathered enough positive evidence to show real promise. The significant intoxi-cating side effects and social stigmas have limited medical use until very recently, but as research continues to hone in on molecular targets, mechanisms, minimizing side effects and further elucidating benefits, we will likely continue to see increases in the wide-spread therapeutic uses of these molecules.

Agronomic Principles for Cannabinoid Production

Cannabinoids are present throughout the plant, but are mostly concentrated in tissues from female flowers, especially in the leaf hairs (trichomes) on the bracts of female flow-ers. Cannabinoids are found at much lower concentrations in root, shoot, and leaf tissues, and are not found in significant concentra-tions in hemp seed, seed oil, or pollen. In the case of optimizing cannabinoid production on a field scale, it is not known if the entire plant would be harvested and processed for canna-binoids, or just the female flowers. Field-scale cannabinoid production could be a case where male plants are totally unwanted. The concen-trations of cannabinoids in male plants is very low compared with female flowers. Also, it is reported anecdotally that unfertilized (un-pol-linated) female flowers tend to produce more cannabinoids than when they are pollinated and allowed to produce seed. As a result, in clonal propagation systems where all female plants are established, attempts would be made to prohibit male plants near the production field. Experiments are underway at UK to quantify the effects of pollination on cannabinoid production in both indoor and outdoor production systems. Early results support the anecdotal premise that unpollinated female flowers produce more cannabinoids than pollinated flowers. However, this work must be repeated to provide a strong level of confidence in the results, and will be published as soon as that can be accomplished.

The biosynthesis of phytocannabinoids is well-described earlier in this chapter. Cannabigerolic acid (CBGA) is the precursor molecule for both THCA and CBDA. Hence, the only difference between plants that we define as hemp or marijuana is the relative abundance of THCA synthase, the enzyme responsible for the reaction of CBA to THCA, which when decarboxylated by time or temperature becomes the intoxicating form of THC. The relative abundance or absence of THCA synthase is genetically controlled. From our understanding of this pathway, it is a reasonable assumption that agronomic inputs that affect the production of either THCA or CBDA would almost certainly have the same effect on the other.

Optimal agronomic protocols for cannabinoid production in field-scale systems have not been well-defined by modern, replicated research methods. Much of what is practiced by cannabinoid producers today is extrapolated from medical and recreational *Cannabis* production systems in U.S. states where it is legal and/or from other countries. Many production practices from these systems (e.g., fertility) pertain mostly to indoor production and not field-scale systems. This section focuses on outdoor or field-scale production. There remains very important and as of yet unanswered questions regarding field-scale production systems for cannabinoids. These include understanding the effects and potential interactions of variety, establishment methods (e.g., direct seeding versus transplanting), pollination, and management decisions including fertility and harvesting, processing, and storage issues. Research is underway to address these questions.

One of the very first crop management decisions in any production system is variety selection. This is especially important with industrial hemp. There are highly significant differences in all yield parameters among hemp varieties grown at specific locations and latitudes. This is due to genetic disposition which defines a variety's capacity to yield high quantities and qualities of fiber, grain, or cannabinoids. While genetically-defined biochemistry is an extremely important consideration, especially regarding cannabinoid production, the photo period sensitivity of *Cannabis* varieties will also be of great importance when considering the latitude where the crop will be produced, for example, varieties derived from northern latitudes may flower too soon when grown at southern latitudes. The opposite is true as well. Varieties derived from southern latitudes may not flower before autumn frosts when grown at northern latitudes.

Field-scale *Cannabis* production solely for the harvest of cannabinoids is a completely new agricultural endeavor in the United States. There are various potential production models currently in practice, but none have not been evaluated by replicated, scientific methods, the results from which could be published in refereed journals. We do not know if cannabinoid yield differences exist between indoor and outdoor growing systems when growing the same variety. Almost all available production knowledge is derived from indoor production systems. This is essentially due to the fact that all production was illicit prior to the legalization of *Cannabis* in the U.S. states of Colorado and Washington in 2012. Prior to legalization, indoor production systems were easier to conceal and control than outdoor systems. Interestingly, a significant portion of *Cannabis* production today is still indoors in states where it is legal. Generally speaking, indoor systems are far more input-rich and hence less sustainable than outdoor production systems. The retail value of the crop today supports these increased inputs, but that will almost certainly change with time as supply meets and then ultimately exceeds demand. There are simply no replicated, refereed, published, scientific studies evaluating outdoor *Cannabis* production systems either alone or in comparison to indoor production systems.

Prior to the establishment of legal industrial hemp production in the United States, cannabinoid production was generally part of a dual-purpose cropping system in Europe, where grain and/or straw were harvested in addition to floral materials for cannabinoid extraction. Since 2014, cannabinoid production systems in the United States have somewhat mirrored marijuana production systems and sometimes other crops (e.g., tomato and/or tobacco). In many modern field-scale systems, clones are produced from cuttings derived from mother plants so as to provide for 100% genetically identical, all female plants. Clones are then rooted, transplanted to the field, and cultured to maturity often without exposure to pollen. Floral material is most often harvested manually, dried for storage, and then ultimately exposed to one or more extraction or purification technologies to derive cannabinoids and other molecules for further formulation and retail sale.

There are very few refereed reports on the effects of pollination on cannabinoid yields. It is a well-known and widely accepted widely anecdotal premise within *Cannabis* culture that unpollinated buds produce significantly more THC than pollinated buds, yet it is surprisingly and extremely difficult to find solid references in the scientific literature to quantify or even validate that hypothesis. ElSohly et al. (1982) conducted an excellent evaluation of the chemical constituents of different forms of *Cannabis*. Included in their analyses were 86 samples of un-pollinated female buds ('sinsemilla'; Spanish for without seed) and 96 samples of pollinated buds ('buds') that contained seed. All seeds and stems were removed from samples prior to analyses. Only ground floral tissues were analyzed. They reported an overall increase in THC concentrations in sinsemilla of two to three times that of the pollinated buds that produced seeds. While the work of ElSohy et al. was not an evaluation of the effects of production practices on cannabinoid yields, the number of samples analyzed does indicate that a relationship almost certainly exists between pollinated and unpollinated flowers. Preliminary research at the University of Kentucky indicated the same levels of increase (approximately two times) in both CBD and THC production in unpollinated versus pollinated plants when growing clones from the same mother in both indoor and outdoor production systems. Additional research is warranted to further delineate the effects of pollination on cannabinoid production.

If a field-scale cannabinoid crop is to remain unpollinated, transplanting female clones is the only reliable and efficient method of production. Rouging male plants from a direct-seeded, dioecious crop has proven to be inefficient and nearly always unsuccessful.

This is also generally true in large scale transplanted systems where the transplants are derived from seed instead of clones.

Current transplanted CBD production models in the United States are establishing between 1000 and 5000 plants per acre; mostly clonal. It is unknown if much denser crops established by direct seeding using conventional drilling equipment will produce competitive or perhaps even optimal cannabinoid yields compared to transplanted clonal systems. Precise plant spacing such as is achieved by transplanting (e.g., tobacco or tomato production models), may have a positive effect on total female flower production per acre and consequently yields of CBD when compared with direct seeding. Precise plant spacing may also have a positive effect on the ease of manual harvesting of female flowers relative to establishment by seeding in rows. There are other potential advantages of transplanting which include: i) use of pre-emergent herbicides with pre-plant cultivation to reduce competition from many weedy species when herbicides become available; ii) use of hemp plants that are at advanced growth stages relative to plants derived from seeding on the same date. An additional advantage of space planting is the increased efficiency of harvesting, drying, and curing the whole plant just as in a tobacco production model. Alternatively, there are CBD production models in the United States harvesting the floral chafe while harvesting the grain or seed from a drilled crop. This is very similar to the European CBD production system. However, there are other issues with this type of production model in the U.S., namely processing and drying of the harvested floral material for storage until extraction. The plant material in Europe is dried in hops driers that are not in use at the same time of year *Cannabis* is harvested. We do not know if the total yields of extractable floral material would be greater in transplanted or direct-seeded (drilled) production models.

Today, there are no automated or mechanical technologies for harvesting unpollinated, whole female flower buds produced in transplanted, space plant systems. Lacking equipment, harvests of whole buds will be conducted manually. Even if mechanical harvesting of floral material existed today, drying the volume of floral material from a field-scale production system prior to storage would still be an imperative issue to address. Large volumes of green plant material (e.g., hay crops) cannot be stored at more than 20% moisture without inciting increased microbial activity leading to rotting, or in some cases, even spontaneous combustion. The same issue would be applicable to hemp floral material if stored above 20% moisture. Optimal moisture content is dependent on how the material is stored; for example, baled versus not. Moisture contents of 12 to 15% would probably be ideal for long term storage of baled hemp floral material. Currently, the infrastructure that's necessary to forcibly dry field-scale volumes of floral material rarely exists in the regions that are cropping hemp for cannabinoid production. Drying floral material in the field as are modern hay crops is certainly an option, but there are concerns about potential cannabinoid yield losses resulting in baling field-dried hemp. It is theorized that trichomes would shatter during raking and baling and consequently would be left in the field. Again, none of these premises have been evaluated by replicated, scientific experiments. How much trichome shattering would occur at 15% moisture? Perhaps we will know in the not-too-distant future.

Infrastructure for and methods of long term storage of green plant material is a serious bottleneck in a mechanically-harvested cannabinoid production system. Research at the University of Kentucky has begun investigations of ensiling hemp. Ensiling plant material is an old method of long-term plant tissue preservation, mostly for animal feeds. It is a very complex biochemical process (Rooke and Hatfield, 2003). Basically, ensiling is a controlled fermentation process providing for preservation by severely limiting microbial activity. Not surprisingly, moisture content for successful ensiling is essentially the opposite of that for long-term dry storage. Plant materials harvested for ensiling should be at or above 20% moisture content.

Initial research indicates that hemp ensiles very well, and that ensiled hemp containing grain (hemp seed) is very nutritious as a potential animal feed. Additional current research is aimed at ensiling un-pollinated female flowers for long-term storage. If the ensiling process does not negatively impact the quality or quantity of cannabinoids within a sample, it could provide for a game-changing process allowing for much more efficient harvesting equipment. Ensiling could be an excellent solution as many farms already understand the ensiling process and have existing equipment and infrastructure to support it. Ensiled floral material could be evenly dried early in the process of preparation for extraction, thus providing for minimal physical disturbance of floral tissues (e.g., shattering of trichomes) during processing.

As noted early in this section, there are no data available from modern, replicated, refereed trials defining agronomic parameters optimizing the yields of cannabinoids from hemp. Conclusions from earlier work are sometimes conflicting. Coffman and Gentner (1975) reported that *Cannabis* plant height was negatively correlated with THC concentrations, suggesting that shorter plants under stress produced THC at higher concentrations than taller, unstressed plants. Additionally, they reported that CBD concentrations were negatively correlated with extractable soil P, but positively correlated with plant N. In a subsequent study, Coffman and Gentner (1977) reported that concentrations of both THC and CBD increased with adequate soil N and P. Caplin et al. (2017) also reported a significant increase in THC production resulting from N fertilization up to 389 mg N L^{-1} in an indoor, liquid application regime. Bócsa et al. (1997) reported a negative correlation between THC production and increasing N fertility, but they measured THC in leaves and not in female flowers. It is interesting to note that all of these studies were conducted using plants grown in pots either in a greenhouse or plant growth chamber. There are no refereed publications reporting on the effects of fertility on cannabinoid yields from plants grown on a field scale. All considered, these results seem to provide an extremely strong impetus for additional studies to elucidate the optimal fertility regimes for maximizing yields of CBD in field-level production systems. We also do not know what effects other macro (e.g., phosphorus and potassium) or micro (e.g., iron and zinc) nutrients may have on cannabinoid yields from field scale systems.

Summary of Agronomic Principles for Cannabinoid Production

Cannabinoid production on a field-scale is a totally new agricultural endeavor. There's just not much known defining optimal production protocols based on replicated, scientific studies. This is true globally. There are currently no official varietal designations or protections for domestically derived germplasm in the United States for high CBD production We do not know if direct-seeded or drilled crops can compete with the yields derived from clonal crops that are transplanted, nor what the optimal plant populations would be in either direct-seeded or transplanted systems. The answer to these questions will be significantly affected by varieties and plant breeding efforts in the future. It is totally feasible that if drying and storage methods and infrastructure are available, new varieties will soon be available that when direct seeded and harvested mechanically will very likely exceed the per acre yields of current clonal transplanted systems. Preventing pollination of clonal plants does appear to increase cannabinoid yields significantly, but will the yields from an increased biomass per acre produced from a direct-seeded model compete with those systems even if the females are pollinated? Can we ensile female hemp flowers for simple, long-term storage and preservation to be extracted months after harvest without significant losses of quality or quantity of cannabinoids? What are the optimal fertility requirements for cannabinoid production? Are there significant interactions between fertility, variety, and production system? These questions and others will necessarily be addressed through scientific research as the cannabinoid industry evolves in the United States.

Literature Cited

Abrams, D.I., P. Couey, S.B. Shade, M.E. Kelly, & N.L. Benowitz. 2011. Cannabinoid-opioid interaction in chronic pain. Clinical pharmacology and therapeutics, 90(6):844–851.

Abrams, D.I., and M. Guzman. 2015. Cannabis in cancer care. Clin. Pharmacol. Ther. 97(6):575–586. doi:10.1002/cpt.108

Adler, J.N., and J.A. Colbert. 2013. Medicinal use of marijuana—polling results. N. Engl. J. Med. 368(22):e30. doi:10.1056/NEJMclde1305159

Agurell, S., M. Halldin, J.E. Lindgren, A. Ohlsson, M. Widman, H. Gillespie, and L. Hollister. 1986. Pharmacokinetics and metabolism of delta 1-tetrahydrocannabinol and other cannabinoids with emphasis on man. Pharmacol. Rev. 38(1):21–43.

Ahmed, A., M.A. Van der marck, G. Van den elsen, M. Olde rikkert. 2015. Cannabinoids in late-onset Alzheimer's disease. Clin. Pharmacol. Ther. 97(6):597–606. doi:10.1002/cpt.117

Amenta, P.S., J.I. Jallo, R.F. Tuma, D.C. Hooper, and M.B. Elliott. 2014. Cannabinoid receptor type-2 stimulation, blockade, and deletion alter the vascular inflammatory responses to traumatic brain injury. J. Neuroinflammation 11:191. doi:10.1186/s12974-014-0191-6

Appendino, G., S. Gibbons, A. Giana, A. Pagani, G. Grassi, M. Stavri, E. Smith, and M.M. Rahman. 2008. Antibacterial cannabinoids from Cannabis sativa: A structure-activity study. J. Nat. Prod. 71(8):1427–1430. doi:10.1021/np8002673

Ashton, C.H. 2001. Pharmacology and effects of cannabis: A brief review. Br. J. Psychiatry 178:101–106. doi:10.1192/bjp.178.2.101

Bailey, J.R., H.C. Cunny, M.G. Paule, and W. Slikker, Jr. 1987. Fetal disposition of delta 9-tetrahydrocannabinol (THC) during late pregnancy in the rhesus monkey. Toxicol. Appl. Pharmacol. 90(2):315–321. doi:10.1016/0041-008X(87)90338-3

Baker, P.B., B.J. Taylor, and T.A. Gough. 1981. The tetrahydrocannabinol and tetrahydrocannabinolic acid content of cannabis products. J. Pharm. Pharmacol. 33(6):369–372. doi:10.1111/j.2042-7158.1981.tb13806.x

Bell, M.R., T.E. D'Ambra, V. Kumar, M.A. Eissenstat, J.L. Herrmann, J.R. Wetzel, D. Rosi, R.E. Philion, S.J. Daum, D.J. Hlasta, R.K. Kullnig, J.H. Ackerman, D.R. Haubrich, D.A. Luttinger, E.R. Baizman, M.S. Miller, and S.J. Ward. 1991. Antinociceptive (aminoalkyl)indoles. J. Med. Chem. 34:1099–1110. doi:10.1021/jm00107a034

Ben Amar, M. 2006. Cannabinoids in medicine: A review of their therapeutic potential. J. Ethnopharmacol. 105(1-2):1–25. doi:10.1016/j.jep.2006.02.001

Bergamaschi, M.M., R.H. Queiroz, M.H. Chagas, D.C. Gomes de Oliveira, B. Spinosa de Martinis, F. Kapczinksi, J. Quevedo, R. Roesler, N. Schroder, A.E. Nardi, R. Martin-Santos, J.E.C. Hallak, and A.W. Zuardi, and J.A.S. Crippa. 2011. Cannabidiol reduces the anxiety induced by simulated public speaking in treatment-naïve social phobia patients. Neuropsychopharmacology 36(6):1219–1226. doi:10.1038/npp.2011.6

Bócsa, I., P. Máthé, and L. Hangyel. 1997. Effect of nitrogen on tetrahydrocannabinol (THC) content in hemp (cannabis sativa L.) leaves at different positions. J. Ind. Hemp 4(2):78–79.

Bonnie, R.J., and C.H. Whitebread. 1970. The forbidden fruit and the tree of knowledge: An inquiry into the legal history of American marijuana prohibition. Va. Law Rev. 56(6):971–1203. doi:10.2307/1071903

Bosy, T.Z., and K.A. Cole. 2000. Consumption and quantitation of delta9-tetrahydrocannabinol in commercially available hemp seed oil products. J. Anal. Toxicol. 24(7):562–566. doi:10.1093/jat/24.7.562

Bouaboula, M., S. Perrachon, L. Milligan, X. Canat, M. Rinaldi-Carmona, M. Portier, F. Barth, B. Calandra, F. Pecceu, J. Lupker, et al. 1997. A selective inverse agonist for central cannabinoid receptor inhibits mitogen-activated protein kinase activation stimulated by insulin or insulin-like growth factor 1. Evidence for a new model of receptor/ligand interactions. J. Biol. Chem. 272:22330–22339.

Burstein, O., N. Shoshan, R. Doron, I. Akirav. 2018. Cannabinoids prevent depressive-like symptoms and alterations in BDNF expression in a rat model of PTSD. Prog. Neuro-Psychopharmacol. Biol. Psychiatry 84(Pt A):129–139. doi:10.1016/j.pnpbp.2018.01.026

BusinessWire. 2016. Ebbu announces groundbreaking scientific research of "entourage effect". BusinessWire 3 July. https://www.businesswire.com/news/home/20160623005867/en/Ebbu-Announces-Groundbreaking-Scientific-Research-Entourage-Effect. (Accessed 3 July 2018).

Busto, U., R. Bendayan, and E.M. Sellers. 1989. Clinical pharmacokinetics of non-opiate abused drugs. Clin. Pharmacokinet. 16(1):1–26. doi:10.2165/00003088-198916010-00001

Caplin, D., M. Dixon, and Y. Zheng. 2017. Optimal rate of organic fertilizer during the vegetative-state for cannabis grown in two coir-based substrates. HortScience 52(9):1307–1312.

Carlini, E.A., I.G. Karniol, P.F. Renault, and C.R. Schuster. 1974. Effects of marihuana in laboratory animals and in man. Br. J. Pharmacol. 50:299–309. doi:10.1111/j.1476-5381.1974.tb08576.x

Carroll, C.B., P.G. Bain, L. Teare, X. Liu, C. Joint, C. Wroath, G. Parkin, P. Fox, D. Wright, J. Hobart, and P. Zajicek. 2004. Cannabis for dyskinesia in Parkinson disease: A randomized double-blind crossover study. Neurology 63(7):1245–1250. doi:10.1212/01.WNL.0000140288.48796.8E

Charuvastra, A., P.D. Friedmann, and M.D. Stein. 2005. Physician attitudes regarding the prescription of medical marijuana. J. Addict. Dis. 24(3):87–93. doi:10.1300/J069v24n03_07

Clark, W.C., M.N. Janal, P. Zeidenberg, and G.G. Nahas. 1981. Effects of moderate and high doses of marihuana on thermal pain: A sensory decision theory analysis. J. Clin. Pharmacol. 21:299S–310S. doi:10.1002/j.1552-4604.1981.tb02608.x

Coffman, C.B., and W.A. Gentner. 1975. Cannabinoid profile and elemental uptake of *cannabis sativa* L. as influenced by soil characteristics. Agron. J. 67:491–497.

Coffman, C.B. and W.A. Gentner. 1977. Responses of greenhouse-grown *cannabis sativa* L. to nitrogen, phosphorus, and potassium. Agron. J. 69:832–836.

Coutts, A.A., and A.A. Izzo. 2004. The gastrointestinal pharmacology of cannabinoids: An update. Curr. Opin. Pharmacol. 4(6):572–579. doi:10.1016/j.coph.2004.05.007

Crippa, J.A., G.N. Derenusson, T.B. Ferrari, L. Wichert-Ana, F.L.S. Duran, R. Martin-Santos, M.V. Simoes, S. Bhattacharyya, P. Fusar-Polli, Z. Atakan, A.S. Filho, M.C. Freitas-Ferrari, P.K. McGuire, A.W. Zuardi, G.F. Busatto, and J.E.C. Hallakl. 2011. Neural basis of anxiolytic effects of cannabidiol (CBD) in generalized social anxiety disorder: A preliminary report. J. Psychopharmacol. (London, U. K.) 25(1):121–130. doi:10.1177/0269881110379283

Davis, M.P. 2016. Cannabinoids for symptom management and cancer therapy: The evidence. J. Natl. Compr. Canc. Netw. 14(7):915–922. doi:10.6004/jnccn.2016.0094

De Petrocellis, L., and V. Di Marzo. 2010. Non-CB1, non-CB2 receptors for endocannabinoids, plant cannabinoids, and synthetic cannabimimetics: Focus on G-protein-coupled receptors and transient receptor potential channels. J. Neuroimmune Pharmacol. 5:103–121. doi:10.1007/s11481-009-9177-z

Devane, W.A., F.A. Dysark, M.R. Johnson, L.S. Melvin, and A.C. Howlett. 1988. Determination and characterization of a cannabinoid receptor in rat brain. Mol. Pharmacol. 34:605–613.

Devinsky, O., J.H. Cross, L. Laux, E. Marsh, I. Miller, R. Nabbout, I.E. Scheffer, E.A. Thiele, and S.M. Wright. 2017. Trial of cannabidiol for drug-resistant seizures in the Dravet syndrome. N. Engl. J. Med. 376(21):2011–2020. doi:10.1056/NEJMoa1611618

Devinsky, O., E. Marsh, D. Friedman, E. Thiele, L. Laux, J. Sullivan, I. Miller, et al. 2016. Cannabidiol in patients with treatment-resistant epilepsy: An open-label interventional trial. Lancet Neurol. 15(3):270–278. doi:10.1016/S1474-4422(15)00379-8

Di Marzo, V., L. De Petrocellis, T. Bisogno, and S. Maurelli. 1995. Pharmacology and physiology of the endogenous cannabimimetic mediator anandamide. J. Drug Dev. Clin. Pract. 7:199–219.

Ellis, G.J., M.A. Mann, B.A. Judson, N.T. Schramm, and A. Taschian. 1985. Excretion patterns of cannabinoid metabolites after last use in a group of chronic users. Clin. Pharmacol. Ther. 38:572–578. doi:10.1038/clpt.1985.226

Elsohly, H.N., C.E. Turner, A.M. Clark, and M.A. Eisohly. 1982. Synthesis and antimicrobial activities of certain cannabichromene and cannabigerol related compounds. J. Pharm. Sci. 71(12):1319–1323.

ElSohly, M.A., and D. Slade. 2005. Chemical constituents of marijuana: The complex mixture of natural cannabinoids. Life Sci. 78(5):539–548. doi:10.1016/j.lfs.2005.09.011

Evans, F.J. 1991. Cannabinoids: The separation of central from peripheral effects on a structural basis. Planta Med. 57(7):S60–S67. doi:10.1055/s-2006-960231

Fellermeier, M., and M.H. Zenk. 1998. Prenylation of olivetolate by a hemp transferase yields cannabigerolic acid, the precursor of tetrahydrocannabinol. FEBS Lett. 427:283–285. doi:10.1016/S0014-5793(98)00450-5

Fellermeier, M., W. Eisenreich, A. Bacher, and M.H. Zenk. 2001. Biosynthesis of cannabinoids. Incorporation experiments with (13)C-labeled glucoses. Eur. J. Biochem. 268(6):1596–1604. doi:10.1046/j.1432-1327.2001.02030.x

Fraser, G.A. 2009. The use of a synthetic cannabinoid in the management of treatment-resistant nightmares in posttraumatic stress disorder (PTSD). CNS Neurosci. Ther. 15(1):84–88. doi:10.1111/j.1755-5949.2008.00071.x

Galli, J.A., R.A. Sawaya, and F.K. Friedenberg. 2011. Cannabinoid hyperemesis syndrome. Curr. Drug Abuse Rev. 4(4):241–249. doi:10.2174/1874473711104040241

Gallily, R., Z. Yekhtin, and L. Hanuš. 2015. Overcoming the bell-shaped dose–response of cannabidiol by using cannabis extract enriched in cannabidiol. Pharmacol. Pharm. 06:75–85. doi:10.4236/pp.2015.62010

Gaoni, Y., and R. Mechoulam. 1964. Isolation, structure, and partial synthesis of an active constituent of hashish. J. Am. Chem. Soc. 86:1646–1647. doi:10.1021/ja01062a046

Gloss, D., and B. Vickrey. 2014. Cannabinoids for epilepsy. The Cochrane database of systematic reviews, 3: CD009270.

Gofshteyn, J.S., A. Wilfong, O. Devinsky, J. Bluvstein, J. Charuta, M.A. Ciliberto, L. Laux, and E.D. Marsh. 2016. Cannabidiol as a potential treatment for febrile infection-related epilepsy syndrome (FIRES) in the acute and chronic phases. J. Child Neurol. 32(1):35–40.

Gorter, R.W., M. Butorac, E. Pulido Cobian, and W. van der Sluis. 2005. Medical use of cannabis in the Netherlands. Neurology 64:917–919. doi:10.1212/01.WNL.0000152845.09088.28

Green, K. 1998. Marijuana smoking vs cannabinoids for glaucoma therapy. Arch. Ophthalmol. 116(11):1433–1437. doi:10.1001/archopht.116.11.1433

Greer, G.R., C.S. Grob, and A.L. Halberstadt. 2014. PTSD symptom reports of patients evaluated for the New Mexico medical cannabis program. J. Psychoactive Drugs 46(1):73–77. doi:10.1080/02791072.2013.873843

Grigoryev, A., S. Savchuk, A. Melnik, N. Moskaleva, J. Dzhurko, M. Ershov, A. Nosyrev, A. Vedenin, B. Izotov, I. Zabirova, and V. Rozhanets. 2011. Chromatography-mass spectrometry studies on the metabolism of synthetic cannabinoids JWH-018 and JWH-073, psychoactive components of smoking mixtures. J. Chromatogr. B Analyt. Technol. Biomed. Life Sci. 879:1126–1136. doi:10.1016/j.jchromb.2011.03.034

Grotenhermen, F. 2003. Pharmacokinetics and pharmacodynamics of cannabinoids. Clin. Pharmacokinet. 42:327–360. doi:10.2165/00003088-200342040-00003

Grotenhermen, F. 2005. Cannabinoids. Curr. Drug Targets CNS Neurol. Disord. 4:507–530. doi:10.2174/156800705774322111

Hampson, A.J., M. Grimaldi, J. Axelrod, and D. Wink. 1998. Cannabidiol and (-)Delta9-tetrahydrocannabinol are neuroprotective antioxidants. Proc. Natl. Acad. Sci. USA 95(14):8268–8273. doi:10.1073/pnas.95.14.8268

Hartman, R.L., T.L. Brown, G. Milavetz, A. Spurgin, D.A. Gorelick, G. Gaffney, and M.A. Huestis. 2015. Controlled cannabis vaporizer administration: Blood and plasma cannabinoids with and without alcohol. Clin. Chem. 61(6):850–869. doi:10.1373/clinchem.2015.238287

Hawks, R.L. 1982. The constituents of cannabis and the disposition and metabolism of cannabinoids. NIDA Res. Monogr. 42:125–137.

Hepler, R.S., and I.R. Frank. 1971. Marihuana smoking and intraocular pressure. JAMA 217:1392. doi:10.1001/jama.1971.03190100074024

Himmi, T., M. Dallaporta, J. Perrin, and J.C. Orsini. 1996. Neuronal responses to delta 9-tetrahydrocannabinol in the solitary tract nucleus. Eur. J. Pharmacol. 312(3):273–279. doi:10.1016/0014-2999(96)00490-6

Hoffmann, D.E., and E. Weber. 2010. Medical marijuana and the law. N. Engl. J. Med. 362(16):1453–1457. doi:10.1056/NEJMp1000695

Holler, J.M., M.L. Smith, S.N. Paul, M.R. Past, and B.D. Paul. 2008. Isomerization of delta-9-THC to delta-8-THC when tested as trifluoroacetyl-, pentafluoropropionyl-, or heptafluorobutyryl- derivatives. Journal of mass spectrometry. J. Mass Spectrom. 43:674–679. doi:10.1002/jms.1375

Howlett, A.C., F. Barth, T.I. Bonner, G. Cabral, P. Casellas, W.A. Devane, C.C. Felder, M. Herkenham, K. Mackie, B.R. Martin, R. Mechoulam, and R.G. Pertwee. 2002. International Union of Pharmacology. XXVII. Classification of cannabinoid receptors. Pharmacol. Rev. 54(2):161–202. doi:10.1124/pr.54.2.161

Järvinen, T., D.W. Pate, and K. Laine. 2002. Cannabinoids in the treatment of glaucoma. Pharmacol. Ther. 95:203–220. doi:10.1016/S0163-7258(02)00259-0

Jain, A.K., J.R. Ryan, F.G. McMahon, and G. Smith. 1981. Evaluation of intramuscular levonantradol and placebo in acute postoperative pain. J. Clin. Pharmacol. 21:320S–326S. doi:10.1002/j.1552-4604.1981.tb02610.x

Jatoi, A., H.E. Windschitl, C.L. Loprinzi, J.A. Sloan, S.R. Dakhil, J.A. Mailliard, S. Pundaleeka, C.G. Kardinal, T.R. Fitch, J.E. Krook, P.J. Novotny, and B. Christensen. 2002. Dronabinol versus megestrol acetate versus combination therapy for cancer-associated anorexia: A North Central cancer treatment group study. J. Clin. Oncol. 20(2):567–573. doi:10.1200/JCO.2002.20.2.567

Johnson, J.R., M. Burnell-Nugent, D. Lossignol, E.D. Ganae-Motan, R. Potts, and M.T. Fallon. 2010. Multicenter, double-blind, randomized, placebo-controlled, parallel-group study of the efficacy, safety, and tolerability of THC:CBD extract and THC extract in patients with intractable cancer-related pain. J. Pain Symptom Manage. 39(2):167–179. doi:10.1016/j.jpainsymman.2009.06.008

Johnson, M.R., and L.S. Melvin. 1986. The discovery of nonclassical cannabinoid analgetics. In: R. Mechoulam, editor, Cannabinoids as therapeutic agents. CRC Press, Boca Raton, FL. p. 121–145.

Jones, R.T. 2002. Cardiovascular system effects of marijuana. J. Clin. Pharmacol. 42:58S–63S. doi:10.1002/j.1552-4604.2002.tb06004.x

Kojoma, M., H. Seki, S. Yoshida, and T. Muranaka. 2006. DNA polymorphisms in the tetrahydrocannabinolic acid (THCA) synthase gene in "drug-type" and "fiber-type" Cannabis sativa L. Forensic Sci. Int. 159(2-3):132–140. doi:10.1016/j.forsciint.2005.07.005

Koppel, B.S., J.C. Brust, T. Fife, J. Bronstein, S. Youssof, G. Gronseth, and D. Gloss. 2014. Systematic review: Efficacy and safety of medical marijuana in selected neurologic disorders: Report of the guideline development subcommittee of the American Academy of Neurology. Neurology 82(17):1556–1563. doi:10.1212/WNL.0000000000000363

Kumar, R.N., W.A. Chambers, and R.G. Pertwee. 2001. Pharmacological actions and therapeutic uses of cannabis and cannabinoids. Anaesthesia 56:1059–1068. doi:10.1046/j.1365-2044.2001.02269.x

Lane, M., C.L. Vogel, J. Ferguson, S. Krasnow, J.L. Saiers, J. Hamm, K. Salva, P.H. Wiernik, C.P. Holroyde, S. Hammill, K. Shepard, and T. Plasse. 1991. Dronabinol and prochlorperazine in combination for treatment of cancer chemotherapy-induced nausea and vomiting. J. Pain Symptom Manage. 6(6):352–359. doi:10.1016/0885-3924(91)90026-Z

LaPoint, J.M. Cannabinoids. In: L.S. Nelson, M. Howland, N.A. Lewin, S.W. Smith, L.R. Goldfrank, and R.S. Hoffman, editors. Goldfrank's toxicologic emergencies. McGraw-Hill, New York, NY. http://accessemergencymedicine.mhmedical.com/content.aspx?bookid=2569&sectionid=210259605. (Accessed 2 June 2019.)

Lee, M. 2012. Smoke signals: A social history of marijuana—Medical, recreational, scientific. Scriber, New York, NY.

Li, H.L. 1974. An archaelogical and historical account of cannabis in China. Econ. Bot. 28:437–448. doi:10.1007/BF02862859

Ligresti, A., A.S. Moriello, K. Starowicz, I. Matias, S. Pisanti, L. De Petrocellis, C. Laezza, G. Portella, M. Bifulco, and V. Di Marzo. 2006. Antitumor activity of plant cannabinoids with emphasis on the effect of cannabidiol on human breast carcinoma. J. Pharmacol. Exp. Ther. 318(3):1375–1387. doi:10.1124/jpet.106.105247

Lotan, I., T.A. Treves, Y. Roditi, and R. Djaldetti. 2014. Cannabis (medical marijuana) treatment for motor and non-motor symptoms of Parkinson disease: An open-label observational study. Clin. Neuropharmacol. 37(2):41–44. doi:10.1097/WNF.0000000000000016

Lutge, E.E., A. Gray, and N. Siegfried. 2013. The medical use of cannabis for reducing morbidity and mortality in patients with HIV/AIDS. Cochrane Database Syst. Rev. (4):CD005175.

Macnab, A., E. Anderson, and L. Susak. 1989. Ingestion of cannabis: A cause of coma in children. Pediatr. Emerg. Care 5:238–239. doi:10.1097/00006565-198912000-00010

Mandolini, G.M., M. Lazzaretti, A. Pigoni, L. Oldani, G. Delvecchio, and P. Brambilla. 2018. Pharmacological properties of cannabidiol in the treatment of psychiatric disorders: A critical overview. Epidemiol. Psychiatr. Sci. 2018:1–9.

Manzanares, J., M. Julian, and A. Carrascosa. 2006. Role of the cannabinoid system in pain control and therapeutic implications for the management of acute and chronic pain episodes. Curr. Neuropharmacol. 4(3):239–257. doi:10.2174/157015906778019527

Martín-Moreno, A.M., D. Reigada, B.G. Ramírez, R. Mechoulam, N. Innamorato, A. Cuadrado, M.L. de Ceballos. 2011. Cannabidiol and other cannabinoids reduce microglial activation in vitro and in vivo: Relevance to Alzheimer's disease. Mol. Pharmacol. 79(6):964–973. doi:10.1124/mol.111.071290

Mathew, R.J., W.H. Wilson, R.E. Coleman, T.G. Turkington, and T.R. DeGrado. 1997. Marijuana intoxication and brain activation in marijuana smokers. Life Sci. 60:2075–2089. doi:10.1016/S0024-3205(97)00195-1

McCallum, N.D., B. Yagen, S. Levy, and R. Mechoulam. 1975. Cannabinol: A rapidly formed metabolite of delta-1- and delta-6-tetrahydrocannabinol. Experientia 31(5):520–521. doi:10.1007/BF01932433

McHugh, D., S.S. Hu, N. Rimmerman, A. Junkat, Z. Vogel, J.M. Walker, and H.B. Bradshaw. 2010. N-arachidonoyl glycine, an abundant endogenous lipid, potently drives directed cellular migration through GPR18, the putative abnormal cannabidiol receptor. BMC Neurosci. 11:44. doi:10.1186/1471-2202-11-44

Mechoulam, R., and L. Hanus. 2000. A historical overview of chemical research on cannabinoids. Chem. Phys. Lipids 108(1–2): 1–13.

Meiri, E., H. Jhangiani, J.J. Vredenburgh, L.M. Barbato, F.J. Carter, H.M. Yang, and V. Baranowski. 2007. Efficacy of dronabinol alone and in combination with ondansetron versus ondansetron alone for delayed chemotherapy-induced nausea and vomiting. Curr. Med. Res. Opin. 23(3):533–543. doi:10.1185/030079907X167525

Melvin, L.S., M.R. Johnson, C.A. Harbert, G.M. Milne, and A. Weissman. 1984. A cannabinoid derived prototypical analgesic. J. Med. Chem. 27:67–71. doi:10.1021/jm00367a013

Mikawa, Y., S. Matsuda, T. Kanagawa, et al. 1997. Ocular activity of topically administered anandamide in the rabbit. Jpn. J. Ophthalmol. 41:217–220. doi:10.1016/S0021-5155(97)00050-6

Mittleman, M.A., R.A. Lewis, M. Maclure, J.B. Sherwood, and J.E. Muller. 2001. Triggering myocardial infarction by marijuana. Circulation 103:2805–2809. doi:10.1161/01.CIR.103.23.2805

Morimoto, S., K. Komatsu, F. Taura, and Y. Shoyama. 1998. Purification and characterization of cannabichromenic acid synthase from Cannabis sativa. Phytochemistry 49:1525–1529. doi:10.1016/S0031-9422(98)00278-7

Munro, S., K.L. Thomas, and M. Abu-Shaar. 1993. Molecular characterization of a peripheral receptor for cannabinoids. Nature 365:61–65. doi:10.1038/365061a0

Nagarkatti, P., R. Pandey, S.A. Rieder, V.L. Hegde, and M. Nagarkatti. 2009. Cannabinoids as novel anti-inflammatory drugs. Future Med. Chem. 1(7):1333–1349. doi:10.4155/fmc.09.93

Nahas, G.G. 1971. Lethal cannabis intoxication. N. Engl. J. Med. 284:792. doi:10.1056/NEJM197104082841417

Narconon International. 2018. History of marijuana. Narconon International, Los Angeles, CA. https://www.narconon.org/drug-information/marijuana-history.html (Accessed 21 June 2018). [2018 is year accessed].

Naveh, N., C. Weissman, S. Muchtar, S. Benita, and R. Mechoulam. 2000. A submicron emulsion of HU-211, a synthetic cannabinoid, reduces intraocular pressure in rabbits. Graefes Arch. Clin. Exp. Ophthalmol. 238:334–338. doi:10.1007/s004170050361

New Frontier. The Cannabis Industry 2017 Annual Report. New Frontier, Denver, CO. https://newfrontierdata.com/annualreport2017/. (Accessed 3 July 2018).

Nguyen, B.M., D. Kim, S. Bricker, F. Bongard, A. Neville, B. Putnam, J. Smith, and D. Plurad. 2014. Effect of marijuana use on outcomes in traumatic brain injury. Am. Surg. 80(10):979–983.

O'Shaugnessy, W.B. 1839. On the preparations of the Indian hemp or gunjah (Cannabis indica): Their effects on the animal system in health, and their utility in the treatment of tetanus and other convulsive diseases. Transactions of Medical and Physical Society of Bengal, p. 421–461.

Perez-Reyes, M., and M.E. Wall. 1982. Presence of delta9-tetrahydrocannabinol in human milk. N. Engl. J. Med. 307:819–820. doi:10.1056/NEJM198209233071311

Pertwee, R.G. 2005. Pharmacological actions of cannabinoids. Handb Exp Pharmacol. 2005(168):1–51.

Pertwee, R.G. 2008. The diverse CB1 and CB2 receptor pharmacology of three plant cannabinoids: Delta9-tetrahydrocannabinol, cannabidiol and delta9-tetrahydrocannabivarin. Br. J. Pharmacol. 153(2):199–215. doi:10.1038/sj.bjp.0707442

Plasse, T.F., R.W. Gorter, S.H. Krasnow, M. Lane, K.V. Shepard, and R.G. Wadleigh. 1991. Recent clinical experience with dronabinol. Pharmacol. Biochem. Behav. 40(3):695–700. doi:10.1016/0091-3057(91)90385-F

Portenoy, R.K., E.D. Ganae-Motan, S. Allende, R. Yanagihara, L. Shaiova, S. Weinstein, R. McQuade, S. Wright and M.T. Fallon. 2012. Nabiximols for opioid-treated cancer patients with poorly-controlled chronic pain: A randomized, placebo-controlled, graded-dose trial. J. Pain 13(5):438–449. doi:10.1016/j.jpain.2012.01.003

Porter, B.E., and C. Jacobson. 2013. Report of a parent survey of cannabidiol-enriched cannabis use in pediatric treatment-resistant epilepsy. Epilepsy Behav. 29(3):574–577. doi:10.1016/j.yebeh.2013.08.037

Prasad, B., M.G. Radulovacki, and D.W. Carley. 2013. Proof of concept trial of dronabinol in obstructive sleep apnea. Front. Psychiatry 4:1. doi:10.3389/fpsyt.2013.00001

Reid, M.J., and L.M. Bornheim. 2001. Cannabinoid-induced alterations in brain disposition of drugs of abuse. Biochem. Pharmacol. 61(11):1357–1367. doi:10.1016/S0006-2952(01)00616-5

Reiner, A., S.A. Heldt, C.S. Presley, N.H. Guley, A.J. Elberger, Y. Deng, L. D'Surney, J.T. Rogers, J. Ferrell, W. Bu, N. Del Mar, M.G. Honig, S.N. Gurley, and B.M. Moore II. 2014. Motor, visual and emotional deficits in mice after closed-head mild traumatic brain injury are alleviated by the novel CB2 inverse agonist SMM-189. Int. J. Mol. Sci. 16(1):758–787. doi:10.3390/ijms16010758

Resstel, L.B., R.F. Tavares, S.F. Lisboa, S.R. Joca, F.M. Corrêa, and F.S. Guimarães. 2009. 5-HT1A receptors are involved in the cannabidiol-induced attenuation of behavioural and cardiovascular responses to acute restraint stress in rats. Br. J. Pharmacol. 156(1):181–188. doi:10.1111/j.1476-5381.2008.00046.x

Rezkalla, S.H., P. Sharma, and R.A. Kloner. 2003. Coronary no-flow and ventricular tachycardia associated with habitual marijuana use. Ann. Emerg. Med. 42:365–369. doi:10.1016/S0196-0644(03)00426-8

Riedel, G., P. Fadda, S. McKillop Smith, R.G. Pertwee, B. Platt, and L. Robinson. 2009. Synthetic and plant-derived cannabinoid receptor antagonists show hypophagic properties in fasted and non-fasted mice. Br. J. Pharmacol. 156:1154–1166. doi:10.1111/j.1476-5381.2008.00107.x

Robson, P., 2001. Therapeutic aspects of cannabis and cannabinoids. British Journal of Psychiatry 178: 107–115. doi:10.1192/bjp.178.2.107

Rooke, J.A., and R.D. Hatfield. 2003. Biochemistry of ensiling. USDA-ARS and University of Nebraska-Lincoln, Lincoln, NE.

Russo, E.B. 2007. History of cannabis and its preparations in saga, science, and sobriquet. Chem. Biodivers. 4:1614–1648. doi:10.1002/cbdv.200790144

Russo, E.B. 2011. Taming THC: Potential cannabis synergy and phytocannabinoid-terpenoid entourage effects. Br. J. Pharmacol. 163(7):1344–1364. doi:10.1111/j.1476-5381.2011.01238.x

Shoyama, Y., M. Yagi, I. Nishioka, and T. Yamauchi, T. 1975. Biosynthesis of cannabinoid acids. Phytochemistry 14:2189–2192. doi:10.1016/S0031-9422(00)91096-3

Sieradzan, K.A., S.H. Fox, M. Hill, J.P. Dick, A.R. Crossman, and J.M. Brotchie. 2001. Cannabinoids reduce levodopa-induced dyskinesia in Parkinson's disease: A pilot study. Neurology 57(11):2108–2111. doi:10.1212/WNL.57.11.2108

Smith, D.E. 1998. Review of the American Medical Association Council on Scientific Affairs Report on medical marijuana. J. Psychoactive Drugs 30:127–136. doi:10.1080/02791072.1998.10399682

Solomon, D. 1968. The marihuana papers. Bobbs-Merril, Indianapolis.

Stockings, E., G. Campbell, W.D. Hall, S. Nielsen, D. Zagic, R. Rahman, B. Murnion, M. Farrell, M. Weier, and L. Degenhardt. 2018. Cannabis and cannabinoids for the treatment of people with chronic non-cancer pain conditions: A systematic review and meta-analysis of controlled and observational studies. Pain 159(10):1932–1954.

Strasser, F., D. Luftner, K. Possinger, G. Ernst, T. Ruhstaller, W. Meissner, Y.D. Ko, M. Schnelle, M. Reif, and T. Cerny. 2006. Comparison of orally administered cannabis extract and delta-9-tetrahydrocannabinol in treating patients with cancer-related anorexia-cachexia syndrome: A multicenter, phase III, randomized, double-blind, placebo-controlled clinical trial from the Cannabis-In-Cachexia-Study-Group. J. Clin. Oncol. 24(21):3394–3400. doi:10.1200/JCO.2005.05.1847

Tashkin, D.P. 2001. Airway effects of marijuana, cocaine, and other inhaled illicit agents. Curr. Opin. Pulm. Med. 7:43–61. doi:10.1097/00063198-200103000-00001

Taura, F., S. Morimoto, and Y. Shoyama. 1995. First direct evidence for the mechanism of D1-tetrahydrocannabinolic acid biosynthesis. J. Am. Chem. Soc. 117:9766–9767. doi:10.1021/ja00143a024

Tramèr, M.R., D. Carroll, F.A. Campbell, D.J. Reynolds, R.A. Moore, and H.J. Mcquay. 2001. Cannabinoids for control of chemotherapy induced nausea and vomiting: Quantitative systematic review. BMJ 323(7303):16–21. doi:10.1136/bmj.323.7303.16

Turner, J.C., J.K. Hemphill, and P.G. Mahlberg. 1978. Quantitative determination of cannabinoids in individual glandular trichomes of Cannabis sativa L. (Cannabaceae). Am. J. Bot. 65:1103–1106. doi:10.1002/j.1537-2197.1978.tb06177.x

United Nations Office on Drugs and Crime (UNODC). 2011. Synthetic cannabinoids in herbal products, Vienna, 2011: 5; see also Hudson, S. Ramsey, J. 'The emergence and analysis of synthetic cannabinoids'. Drug Test. Anal. 3:466–478.

Wallace, E.A., S.E. Andrews, C.L. Garmany, and M.J. Jelley. 2011. Cannabinoid hyperemesis syndrome: Literature review and proposed diagnosis and treatment algorithm. South. Med. J. 104:659–664. doi:10.1097/SMJ.0b013e3182297d57

Ware, M.A., M.A. Fitzcharles, L. Joseph, and Y. Shir. 2010. The effects of nabilone on sleep in fibromyalgia: Results of a randomized controlled trial. Anesth. Analg. 110(2):604–610. doi:10.1213/ANE.0b013e3181c76f70

Watanabe, K., S. Yamaori, T. Funahashi, T. Kimura, and I. Yamamoto. 2007. Cytochrome P450 enzymes involved in the metabolism of tetrahydrocannabinols and cannabinol by human hepatic microsomes. Life Sci. 80(15):1415–1419. doi:10.1016/j.lfs.2006.12.032

Whiting, P.F., R.F. Wolff, S. Deshpande, M. Di Nisio, S. Duffy, A.V. Hernandez, J.C. Keurentjes, S. Lang, K. Misso, S. Ryder, S. Schmidlkofer, M. Westwood, and J. Kleijnen . 2015. Cannabinoids for medical use: A systematic review and meta-analysis. JAMA 313(24):2456–2473. doi:10.1001/jama.2015.6358

Wiley, J. L., Marusich, J. A., Huffman, J. W., Balster, R. L., & Thomas, B. F. 2011. Hijacking of basic research: The case of synthetic cannabinoids. Methods report (RTI Press), 2011, 17971. doi:10.3768/rtipress.2011.op.0007.1111

Williams, S.J., J.P. Hartley, and J.D. Graham. 1976. Bronchodilator effect of delta1-tetrahydrocannabinol administered by aerosol of asthmatic patients. Thorax 31:720–723. doi:10.1136/thx.31.6.720

Williamson, E.M., and F.J. Evans. 2000. Cannabinoids in clinical practice. Drugs 60(6):1303–1314. doi:10.2165/00003495-200060060-00005

Wills, S. 2005. Drugs of abuse. 2nd ed. Pharmaceutical Press, London, U.K.

Wu, T.C., D.P. Tashkin, B. Djahed, and J.E. Rose. 1988. Pulmonary hazards of smoking marijuana as compared with tobacco. N. Engl. J. Med. 318:347–351. doi:10.1056/NEJM198802113180603

Yamaori, S., K. Koeda, M. Kushihara, Y. Hada, I. Yamamoto, and K. Watanabe. 2012. Comparison in the in vitro inhibitory effects of major phytocannabinoids and polycyclic aromatic hydrocarbons contained in marijuana smoke on cytochrome P450 2C9 activity. Drug Metab. Pharmacokinet. 27:294–300. doi:10.2133/dmpk.DMPK-11-RG-107

Yamaori, S., M. Kushihara, I. Yamamoto, and K. Watanabe. 2010. Characterization of major phytocannabinoids, cannabidiol and cannabinol, as isoform-selective and potent inhibitors of human CYP1 enzymes. Biochem. Pharmacol. 79(11):1691–1698. doi:10.1016/j.bcp.2010.01.028

Yamaori, S., Y. Okamoto, I. Yamamoto, and K. Watanabe. 2012. Cannabidiol, a major phytocannabinoid, as a potent atypical inhibitor for CYP2D6. Drug Metab. Dispos. 39(11):2049–2056. doi:10.1124/dmd.111.041384

Zeese, K.B. 1999. History of medical marijuana policy in US. Int. J. Drug Policy 10(4):319–328. doi:10.1016/S0955-3959(99)00031-6

Zhang, H., D.M. Lund, H.A. Ciccone, W.D. Staatz, M.M. Ibrahim, T.M. Largent-Milnes, H.H. Seltzman, I. Spigelman, and T.W. Vanderah. 2018. A peripherally restricted cannabinoid 1 receptor agonist as a novel analgesic in cancer-induced bone pain. Pain 159(9):1814–1823.

Zuardi, A.W. 2006. History of cannabis as a medicine: A review. Rev. Bras. Psiquiatr. 28:153–157. doi:10.1590/S1516-44462006000200015

Zuardi, A.W., R.A. Cosme, F.G. Graeff, and F.S. Guimarães. 1993. Effects of ipsapirone and cannabidiol on human experimental anxiety. J. Psychopharmacol. (London, U. K.) 7(1, Suppl)82–88. doi:10.1177/026988119300700112

Zullino, D., D. Delessert, C. Eap, M. Preisig, and P. Baumann. 2002. Tobacco and cannabis smoking cessation can lead to intoxication with clozapine or olanzapine. Int. Clin. Psychopharmacol. 17(3):141–143. doi:10.1097/00004850-200205000-00008

Photo by Marcia O'Connor

Chapter 6: Hemp Genetics and Genomics

Brian Campbell, Dong Zhang, and John K. McKay*

Introduction

Cannabis sativa L. is an economically important crop that has been surrounded by controversy over the last 100 years. Despite its widespread use as an intoxicant and an industrial crop, governments worldwide have struggled to appropriately regulate Cannabis use and production. The lack of uniformity in Cannabis law, both spatially and temporally, has made research on this plant difficult through traditional channels. With increasing public support for medical marijuana and growing interest in applications for industrial hemp, laws are changing and doors are opening providing for long-overdue research. The purpose of this chapter is to review the current body of knowledge relative to Cannabis genetics, specifically regarding speciation and classification, evolutionary origin, genetics of industrial hemp breeding, genetic diversity and population structure of the species, and Cannabis genomics, as well as a look at important areas of genetics and genomics research for the future of industrial hemp.

The origin and the taxonomy of Cannabis is surrounded by uncertainty; academics debate questions of both where its evolutionary roots lie, as well as whether or not the diversity observed in Cannabis warrants distinguishing common types as separate species or subspecies. Eurasia has been proposed as its evolutionary center of origin, with more specific recommendations of central Asia, in the Himalayas, or possibly two distinct centers of Hindustani and European-Siberian origin (Clarke and Merlin, 2016; Hillig, 2005; Andre et al., 2016). As defined in previous chapters, Cannabis has been utilized since ancient times. Archeological findings date the use of hemp rope to 10,000 years ago in Taiwan (Laursen, 2015). This early human use led to vectoring of seed thousands of years ago (Hillig, 2005) that has made clearly distinguishing evolutionary origins by genetic analyses difficult, if not practically impossible.

The diverse morphology of *C. sativa*, combined with a multitude of uses, has caused no shortage of confusion over the classification of this species. The debate of whether or not Cannabis is a single species began long ago. The species was first labeled *Cannabis sativa* by Carl Linnaeus in 1753 (Watts, 2006). This monospecific viewpoint was challenged in 1785 when Jean-Baptiste Lamarck found that some

---

B. Campbell, D. Zhang, and J.K. McKay, Colorado State University, Fort McCollins, CO 80523. *Corresponding author (jkmckay@colostate.edu)

doi:10.2134/industrialhemp.c6
Industrial Hemp as a Modern Commodity Crop. D.W. Williams, editor.

Cannabis specimens from India exhibited distinctly different morphology from those described by Linnaeus and created a new classification, *Cannabis indica* (Watts, 2006). Sativa samples were tall, with long inter-modal spacing, and narrow leaflets, while indica-types are shorter and bushier plants with wide leaflets. Although several other classifications have been proposed, the only generally accepted third possible species was proposed in 1924 by Dmitri Janischevsky as *Cannabis ruderalis* (Hillig and Mahlberg, 2004). This newest classification of Cannabis was created to describe Russian samples that did not exhibit the same "domestication syndrome" traits as indica and sativa samples and were essentially small, wild (ruderal; or generally occurring in regularly disturbed ecosystems) plants (Small and Cronquist, 1976). Since there are no reproductive barriers between these three types of Cannabis, we consider this as a single species, but the debate between the "splitters" and the "lumpers" remains very active (Watts, 2006).

Chromosome Number and Ploidy Level

Although the taxonomy of Cannabis is debated, it is agreed that plants in the species are diploid and have nine pairs of autosomes, as well as a pair of sex chromosomes that are cytogenetically heteromorphic in male plants (XY system), but homomorphic in female and monoecious plants (XX system) (Divashuk et al., 2014; Faux et al., 2016; Razumova et al., 2016). However, the roles of Y chromosomes and X-to-autosome ratio in the sex determination system are still open questions in Cannabis (Sakamoto et al., 1998; Ming et al., 2011; Faux et al., 2014; reviewed by Vergara et al., 2016). Using DAPI/C-banding staining and FISH, Divashuk et al. (2014) demonstrated the karyotype in dioecious hemp. The cytogenetic study showed that all chromosomes appear to be submetacentric and metacentric (meaning the centromere is located below or near the center of the chromosome, respectively. The location of the centromere assists in definitions of karyotypes and provides for appropriate descriptions of the locations of genes on chromosomes) and the Y chromosome is larger than the X chromosome (Divashuk et al., 2014). In contrast, the Y chromosome was reported to be shorter than X chromosome in male samples of the closely related species,

hops, *Humulus lupulus* (Karlov et al., 2003). The genetic degeneration of Y chromosome after divergence of the ancestors of Cannabis and Humulus deserves further investigation.

Genetics of Hemp Breeding
Breeding Targets

There are thousands of potential products that can be made from hemp, but breeding objectives generally fall within three main categories: fiber, seed, and, more recently, secondary metabolite production (non-THC cannabinoids and terpenes) (Salentijn et al., 2015). Due to this diversity of end uses, hemp breeding can progress in many different directions depending on the goals of the individual breeding program. It is important, as in any breeding program, to have clear goals that align with the production of specific end products.

Hemp's use as a durable fiber dates back thousands of years and archeological evidence shows that it was one of the first fiber plants domesticated by humans (Lynch et al., 2016). As such, improving fiber yield and quality have been primary breeding goals for hemp. Length of vegetation period is directly correlated to fiber yield, so selection based on this trait allowed for steady improvement of stem biomass in early cultivars (Ranalli, 2004). The proportion of bast fiber content to biomass, however, is a more complex trait to improve and little improvement was made until the Bredemann method was employed starting in 1942, which used bast fiber content as a primary criterion for selecting males (Salentijn et al., 2015). This in vivo method involved splitting the main stem to measure bast fiber content on living plants and only allowing males with the highest bast fiber content to flower (Ranalli, 2004). This enabled increased genetic gain and created plants with three times higher bast fiber content over the next 30 yr (Ranalli, 2004). In 1953, Jakobey noted a negative correlation between bast fiber content and stem weight, so he developed a technique called the "normal axis" method to identify plants that broke pattern with the common correlation (Bócsa, 1999). Adding to this work, Horkay (1982) found that there was a strong negative relationship between bast fiber content and fiber quality, where increases in bast fiber were almost entirely secondary fibers of lower quality. After this period, however, breeding for fiber quality was no longer a priority because of the advent of new fiber processing techniques, such as steam explosion processing and ultrasonic refining (Bócsa, 1999).

Hemp's historic breeding efforts have largely focused on increasing fiber yield, but its potential as an oilseed crop is also being considered. Traditional dioecious hemp is grown as a seed crop, but these cultivars often exhibit significant variation and only produce seed on half of the plants. However, intersex plants are a common occurrence in dioecious hemp and can be selected for, with ideal stabilization of sex ratios within six to eight generations (Bócsa, 1999). This creates a crop that has higher uniformity than dioecious types, and they have become common for oilseed or dual-purpose (seed and fiber) cultivars, although many dioecious oilseed varieties are utilized as well (Salentijn et al., 2015). The first true oilseed variety, "FIN-314", was developed in Finland using germplasm from the Vavilov Research Institute Gene Bank and was put into production in Canada in 1998 (Ranalli, 2004) after bans on hemp production were lifted (Salentijn et al., 2015). Canadian farmers in particular have turned to hemp as an alternative oilseed crop. With the support of government programs and a preexisting oilseed infrastructure, oilseed hemp production and breeding have flourished in Canada, with 39 Canadian approved oilseed or dual-purpose cultivars listed as of 2014 (Salentijn et al., 2015).

While breeding for fiber and seed traits began long ago, a much newer breeding target has emerged in the form of cannabinoid profiles. The practical side of breeding for cannabinoid content lies in the legal restrictions on THC content ($< 0.3\%$ dry weight worldwide, $< 0.2\%$ in Europe). Any breeder that sells seed needs to be entirely sure that their variety is in compliance with standards in the production region. There is also current interest in medical hemp, which is harvested for Cannabis' other primary cannabinoid, cannabidiol or CBD. Cannabidiol is one of the major constituents that makes Cannabis medicine, with reported benefits as an antiepileptic, anticonvulsant, neuroprotectant, antioxidant, and as an antianxiety and anti-inflammatory agent (Devinsky et al., 2014). This has led to a boom in breeding high-CBD varieties of hemp. There are both qualitative and quantitative aspects to breeding for chemical phenotype (chemotype) and cannabinoid content, which will be discussed more extensively later. In Colorado and other states with legal marijuana it has become common to cross hemp varieties with marijuana strains to produce a predominantly CBD chemotype with high levels of overall cannabinoid production. The resulting progeny often has unstable THC levels and requires extensive breeding for uniformity and stability before seed can reliably produce a hemp phenotype and be sold as industrial hemp. Most CBD farmers are circumventing this latter restriction by planting tested clones in the field, but this approach is cost and labor intensive and ultimately not scalable or sustainable in the same way that it is possible to produce crops from seed.

Along with these three major breeding objectives, hemp fills many specialty niche roles, for which breeding will be integral. If industries develop around the use of hemp as paper, concrete, composites, textiles, specialty foods, bio-plastics, and more, industry-driven breeding of locally adapted cultivars that maximize specific plant components will become increasingly important (Salentijn et al., 2015).

Breeding Methods

Hemp is a naturally cross-pollinated crop, which, in the absence of strict selection, maintains high levels of natural genetic variation and heterozygosity within populations (Salentijn et al., 2015). As a result, most available hemp cultivars are populations that exhibit phenotypic variation. This can be a challenge for farmers, for instance, when plants are different heights, harvesting grain heads with a combine is problematic. Differences in maturity times can also result in seed loss. Because many hemp varieties were initially bred for fiber (and harvested prior to maturity), it is unclear how much effort was put into breeding for uniform height and reproductive maturity.

Historically, the most common approach to hemp breeding has been recurrent mass selection, where each generation's plants or seeds are selected to create the next generation based on a predetermined trait threshold (Ranalli, 2004). This approach has created many productive cultivars such as Bolognese, Toscana and YunMa 1 by improving landrace varieties, but is limited by the intensity of selection and the heritability of the target traits (Salentijn et al., 2015; Hennink, 1994). Hennink (1994) reported that no studies had even reported estimates of heritability and it appears that maximizing response to selection has not traditionally been a primary goal for hemp breeders.

Despite a generally primitive approach to breeding, individual breeders recognized the potential of exploiting heterosis relatively early. Dewey (1927) is credited with creating the first intervarietal, or synthetic, hybrid by crossing Kymington and Ferrara. The resulting F1 hybrid, Ferraramington, had excellent fiber characteristics and was one of several successful cultivars developed by Dewey (Bócsa, 1999). Unfortunately, all of Dewey's germplasm was lost due to Cannabis prohibition in the United States (Ranalli, 2004). Eventually, crossing varieties became more common as a way to generate new genetic variation in breeding populations and resulted in improved cultivars like the Chinese varieties YunMa 2 and YunMa 4 (Ranalli, 2004; Salentijn et al., 2015).

To maximize heterotic response, it is necessary to first develop inbred lines which are subsequently crossed to make true hybrid cultivars (Bernardo, 2002). One common difficulty in producing hybrids is pollen control, but utilizing self-fertilization in monoecious or subdioecious hemp can produce all female progeny which can act as a proxy for a male sterility system (Salentijn et al., 2015). This method has been used in Hungary and China to produce both single and double cross hybrids that greatly outperformed the parents (Salentijn et al., 2015; Bócsa, 1999). Despite these successes, true hybrid varieties of hemp are still relatively rare and there will need to be a paradigm shift in hemp breeding to realize gains similar to what was achieved in maize throughout the 1900s.

Recent Cannabis Genetic Studies

Genetic Basis for Production of Secondary Metabolites

Due to the previously restricted ability to grow and handle Cannabis plants, research of Cannabis genetics and the development of genetic resources lags far behind other economically important crops. With the recent relaxation of these restrictions, a corresponding increase in all types of Cannabis research has emerged, particularly regarding industrial hemp. Research on the genetic basis of many traits has begun in the last decade, with a heavy initial focus on the genetics of cannabinoid production, as well as sex determination and agronomic traits (e.g., fiber quality).

Since the distinction between marijuana and hemp depends (legally) on the level of THC found in plant material, it has become of primary importance to understand the genetics of cannabinoid biosynthesis. Potential medical uses of Cannabis have also generated a significant amount of interest in THC's non-intoxicating isomer, CBD, as a pharmacological compound to treat a range of ailments from epilepsy to anxiety (van Bakel et al., 2011; Grotenhermen and Müller-Vahl, 2016, Felberbaum and Walsh, 2018). For many years it was thought that cannabidiolic acid (CBDA) was the direct precursor to tetrahydrocannabinolic acid (THCA) (Mechoulam et al., 1970; Shoyama et al., 1975). However the correct biochemistry of this process was elucidated more recently when it was discovered that cannabigerolic acid (CBGA) acts as a precursor to multiple compounds and produces either THCA or CBDA via enzymatic conversion with THCA synthase or CBDA synthase (Taura et al., 1995).

Cannabis has been described as having three common chemotypes distinguished by cannabinoid ratios: high THC/low CBD (marijuana), low THC/high CBD (hemp), and an intermediate ratio of the two compounds (hybrid) (de Meijer et al., 2003). At first glance, the genetics of these categories seem straightforward. A single, codominant locus (B) appears to establish the chemotype with BT and BD alleles producing predominantly THC or CBD, respectively (de Meijer et al., 2003). It was also reported that other rare alleles, BC and BO, create rare chemotypes that produce mainly cannabichromenic acid (CBCA) and a non-functional, cannabinoid-free phenotype (de Meijer et al., 2003). In the same study, de Meijer et al. (2009) proposed that the BO allele may actually be a linked second locus (O) where homozygous O/O combinations produce a normal range of cannabinoids, heterozygous O/o combinations severely reduce cannabinoid production, and homozygous recessive o/o combinations produce a cannabinoid-free phenotype. The BO allele or o/o recessive genotype is particularly interesting as it represents a mechanism for creating industrial hemp cultivars that have nil levels of THC and can be easily utilized via molecular marker-assisted selection.

Weiblen et al. (2015) performed a quantitative trait locus (QTL) study which supported the single locus model of chemotype inheritance, but the distribution of cannabinoid synthase homologs in their mapping population indicated that two or more tightly linked loci could be controlling the trait, an idea initially proposed by de Meijer et al. (2009) regarding rare chemotypes and van Bakel et al. (2011) specifically regarding THCA/CBDA production. Despite previous efforts to categorize cannabinoid production as a qualitative trait, in reality the quantity of cannabinoids present in Cannabis flowers has proven to be a quantitative, polygenic trait. Weiblen et al. (2015) found significantly different quantities of cannabinoids in their study, with marijuana-type Cannabis averaging 4.5 times total cannabinoid levels compared to hemp. However, the small population size (N = 62) did not allow for detection of any significant QTL for cannabinoid quantity. The linkage map reported in this study was created from an F2 population, derived from a cross between a single staminate hemp plant (Carmen) and a single pistillate marijuana plant (Skunk #1) and consists of 9 linkage groups, including 103 AFLP and 16 SSR markers, spaced 6.10 cM on average (Weiblen et al., 2015). Ultimately, the study detected only one significant QTL for qualitative chemotype characterization and one putative QTL for log THCA or CBDA content, located in two separate linkage groups. This supports the idea of separate loci affecting cannabinoid type and content, but the authors acknowledge an inability to detect sufficient QTL to fully characterize the

genetic architecture of cannabinoid production and expect that with higher map density and larger populations it will be possible to detect more QTL and a tenth linkage group will emerge to properly reconcile the linkage map with known chromosome number (Weiblen et al., 2015).

Terpenoid Production

Terpenoids, contributing to the scent and taste of Cannabis and commonly called essential oils, are a source of interest as phytotherapeutic agents, as well as for their hypothesized nonadditive interactions with cannabinoids (Russo, 2011). These compounds are produced in terpene-rich resin, which is mainly synthesized and accumulated in glandular trichomes of female inflorescences in Cannabis (Booth et al., 2017). To date, over 100 terpenoids have been identified in Cannabis, prompting questions of both how these compounds are produced and what possible uses they could fulfill (Andre et al., 2016).

The first study investigating the genetics of terpene synthesis (Booth et al., 2017) showed that transcripts associated with terpene biosynthesis are expressed in glandular trichomes more than in non-resin producing tissues, agreeing with chemical analyses of these tissues. Genomic and transcriptomic data from the hemp variety 'Finola' enabled the identification of nine Cannabis terpene synthases (CsTPS) that account for the majority of terpene production, with the exception of terpinolene which proved elusive (Booth et al., 2017). Similar to cannabinoid production, it appears that quantity of terpenes produced is polygenic and involves the production of competitive enzymes (Booth et al., 2017). Due to an intense interest in characterizing important pharmaceutical interactions in medical marijuana and an emerging interest in hemp as a source of medical and wellness products, research in this area is likely to expand rapidly in the near future.

Sex Expression in Hemp

Although Cannabis is mainly dioecious, monoecious plants are often observed in natural populations and can be intentionally induced via treatment with chemicals or environmental stress (Mohan Ram and Sett, 1982). These monoecious plants lack a Y chromosome, but are still able to produce staminate inflorescences. One interesting aspect of Cannabis is that "sex expression" in monoecious plants has been defined as a quantitative trait rather than a binary trait. A recent study (Faux et al., 2016) quantified sex expression in three hemp F1 populations by the ratio of female and male flowers. Faux et al. (2016) utilized 71 AFLP markers to identify 5 QTL in each of three maps, and showed genetic correspondence of QTL across three maps. However, the study provided relatively low mapping resolution for sex expression due to the low number of markers.

Genetics of Agronomic Traits

There is an emerging picture of the genetics behind the production of secondary compounds and sex expression in Cannabis, but other traits remain unexplored. Although agronomic performance of hemp has been relatively well characterized, exploration of genetics behind major agronomic traits is just beginning. For hemp breeding and production to advance, it is necessary to understand how major quantitative traits are controlled.

Despite an historical breeding focus on fiber quality and quantity, the genetics of these traits are poorly understood. One initial study on fiber quality by van den Broeck et al. (2008) explored the molecular processes underlying cell wall synthesis to lay the groundwork for manipulating content of cellulose and lignin in hemp stem tissue. The authors looked at genes that were differentially expressed in the bast and hurd fibers using a cDNA microarray and found 110 clones with higher expression in bast tissue and 178 clones more highly expressed in hurd tissue. The genes preferentially expressed in the bast tissue were, expectedly, many genes associated with photosynthesis, chlorophyll, and chloroplast production, as well as arabinogalactan proteins. Most of the genes more highly expressed in the hurd tissue were directly related to enzymatic conversion of fructose-6-phosphate to various forms of lignin (van den Broeck et al., 2008). This is relatively unsurprising since the core is the woody section of the stem, but is an important characteristic. For instance, when using hemp fiber for making composite materials, lignin can function as a useful binder, whereas the same compounds lower the quality for textile applications by adding

undesired stiffness (van den Broeck et al., 2008). This study provides information about genes and gene families that are important to biosynthesis of commercially relevant traits, however, utilizing this information is difficult without further study of the degree of impact of individual genes or haplotypes.

Hemp has a long history as a fiber crop, but hemp grain (seed) has been utilized for at least 6000 years as well (Li, 1973). One relevant area of study that has been explored in hemp seed is the genetics of fatty acid production. Hemp seed contains over 80% polyunsaturated fatty acids, with a desirable ratio of linoleic acid and α-linolenic acid, making it an common source of oil and protein for human and animal nutrition dating back to Neolithic times (Li, 1973; Bielecka et al., 2014). In a first of its kind study, Bielecka et al. (2014) created a TILLING population (Targeting Induced Local Lesions in Genomes, Till et al., 2006) of industrial hemp from the oilseed cultivar Finola. This reverse genetics approach, which induces point mutations throughout the genome, allows researchers to observe altered phenotypes in mutant progeny and determine which gene sequences changed to produce these aberrant phenotypes. This particular study focused on D12 and D15 desaturase genes by comparing expressed sequence tags (ESTs) that showed homology to known desaturase genes. This approach identified 12 genes with membrane-bound expression in the FAD2, FAD3, and Δ6/Δ8 sphingo-lipid families and five genes for soluble Δ9 stearoyl-ACP desaturases. Utilizing M2 plants with mutations in these genes, function of these oil metabolism genes was confirmed and a pathway was laid to produce specialized oil profiles in hemp, such as high-oleic hemp (Bielecka et al., 2014). This can have important commercial applications as oilseed varieties of hemp can be used for specific and unrelated end-uses like human consumption or production of biofuel.

The genetics of other important agronomic traits such as seed yield, biomass production, crop uniformity, photoperiod sensitivity, and flowering time have no published studies at the time of this writing. However, the Bielecka et al. (2014) study showed that despite a lack of major genetic resources such as an annotated and anchored genome in Cannabis, it is possible to use homology with other well-studied crops as a shortcut to understanding gene function. The availability of modern tools like affordable Next Generation Sequencing will help allow Cannabis researchers to rapidly catch up to other major crops in the coming decades.

MultiHemp

The TILLING study by Bielecka et al. (2014) was partially funded by the EU Framework Program 7, MultiHemp. MultiHemp is the first major government funded Cannabis research initiative. The program ran from September 2012 to February 2017 with the goal of using "cutting-edge genomic approaches to achieve rapid targeted improvements in hemp productivity and raw material quality for end-user requirements, whilst also advancing scientific understanding of gene-to-trait relationships in this crop" (MultiHemp, 2017). The scope of the project was expansive and included engineering for harvest and processing, hemp agronomy, crop modeling, and genetics/genomics. The Bielecka et al. (2014) study is the first genetics paper to be published from this project, but more genetics projects are underway describing the first Genome-wide Association Study (GWAS) and Heteroduplex mapping in hemp (Multi-Hemp, 2017). This project was an important step in breaking long-held stereotypes about Cannabis. If the United States is to properly contribute to hemp research, it is important for federal granting agencies like the United States Department of Agriculture (USDA) to rapidly create a path for publicly funded hemp research. Additionally, a permanent change in law regarding Cannabis and the distinction between marijuana and hemp would allow researchers to undertake projects without overly burdensome regulation.

Genetics and Genomic Diversity

Reference Genome and Transcriptomes

To determine gene function in any species and understand the relationships between genes and haplotypes with phenotype, an accurate reference assembly is essential (Stemple, 2013). The marijuana strain Purple Kush (PK) was the first published genome in Cannabis, using a combination of Illumina and Roche 454 sequencing with ~130X coverage of the estimated ~820 Mb haploid genome. De novo assembly generated 136,290 scaffolds with a total size of 786.6 Mb, accounting for approximately ~96% of the estimated haploid genome size (van Bakel et al., 2011). However, the genome coverage could be overestimated due to high proportion of redundant scaffolds of homologous regions with high heterozygosity rates (Vergara et al., 2016). Ongoing efforts to accurately assemble and annotate the genome are necessary to more clearly establish full genomic coverage.

A total of 30,074 transcript isoforms were constructed from the transcriptome assembly of PK, in which 83% have homologous counterparts in other plants. The remaining 17% may represent some unique gene models in Cannabis, but also likely represents assembly error and erroneous gene model prediction. Characterization of the transcriptome was paralleled by identification of differential gene expression in root, stem, shoot and three flowering stages. The expression profiles exhibit similar patterns in the six tissues because of widespread expression of photosynthetic processes and primary metabolic pathways in the plants (van Bakel et al., 2011). The authors also explored expression of THCA synthase and CBDA synthase and showed that they are expressed in opposite ways in the marijuana type (PK) and the hemp type (Finola), supporting that qualitative aspects of cannabinoid production are primarily determined by the presence or absence of these enzymes.

The released draft genomes and transcriptomes of marijuana types provide references for genetic variant detection and accelerate progress in genetic mapping and relating Cannabis genes to their functions. Additional genomic resources are being developed and are well reviewed by Vergara et al. (2016). These forthcoming resources will help answer questions such as: what is the content and function of repetitive regions, is there any evidence of ancestral whole-genome duplications, and what are the over- and under-represented gene families? The abundance of repetitive sequences and level of heterozygosity represent challenges in making a chromosome-scale assembly in Cannabis, which could be mitigated with targeted approaches utilizing inbred lines and structured populations. It is also important to characterize sex chromosomes to clarify current pseudo-autosomes, male-specific loci, and the fate and consequences of genes on sex chromosomes.

Comparative Genomics

Genomic resources for Cannabis are relatively sparse compared with model species such as *Arabidopsis thaliana*, so utilizing comparative genomics as a natural extension of Cannabis genomics research will answer questions regarding how Cannabis gene function is both similar to and different from other plant species. Due to strong preservation of homeologous regions, "translation genomics" has been a successful approach for cross-utilization of genetic knowledge of closely related species (Paterson, 1995; Kim et al., 2012). Therefore, an obvious starting point for this approach is to utilize species most closely related to Cannabis. Using four plastid loci (atpB-rbcL, rbcL, rps16 and trnL-trnF), a molecular phylogenetic study confirmed the close genetic relationship between Humulus and Cannabis (Yang et al., 2013), two genera in Cannabaceae family) which diverged around 21 to 27.8 million years ago (Divashuk et al., 2014; Laursen, 2015) (See Fig. 1). The group shows variation with regard to genome size and chromosome numbers among *C. sativa* (male: ~0.84Gb, 2n = 20; female: ~0.81Gb, 2n = 20), *H. japonicus* (male: ~1.7Gb, 2n = 17; female: 2n = 16) (the primitive type of Humulus) and H. lupulus (male: ~2.9Gb, 2n = 20; female: ~2.57Gb, 2n = 20) (van Bakel et al., 2011; Divashuk et al., 2014; Natsume et al., 2014). Understanding the patterns of evolution of genome size and structure among the members in Cannabaceae provides clues about the path of speciation and selection, and the fates of gene families, especially for sex expression.

The prevalence of atypical meiotic configuration, such as translocation heterozygosity, has been implicated in Humulus (Sinoto , 1929; Neve, 1958; Haunold, 1991; Shephard et al., 2000; Zhang et al., 2016). The findings shed light on questions on the unusual transmission genetics and phenotypic variation in hops, yet the abnormal meiotic events have not been reported in cytogenetic studies in Cannabis (Divashuk et al., 2014; Razumova et al., 2016). Due to a shared genetic origin with Humulus species, however, the possibility of atypical meiotic configuration may not automatically be ruled out in Cannabis.

Cannabis and Humulus are frequently characterized by their secondary metabolite systems, producing a variety of chemical compounds contributing to plant growth and human uses. One example of convergent breeding targets can be found in the selection of terpene profiles, which is commercially relevant for both and likely have similar genetic bases. Although this approach has only begun to be explored in Cannabis, important information can be gleaned in this manner, both by comparison with Humulus species as well as more distantly related but better characterized species like *A. thaliana* or maize (*Zea mays* L.).

Genetic Diversity and Population Structure

Debates over the speciation and classification of types of Cannabis are largely rooted in the phenotypic diversity that is apparent in the species. This has led to questions about what the total genetic diversity of Cannabis encompasses and how it is possible to understand and classify this diversity. It is also important to understand genetic patterns in the species that will allow variety and cultivar identification, purity inspection, lineage, and characterization of drug (or medicinal) and non-drug (or non-medicinal) strains. Identification can be addressed by a combination of morphology, chemistry, and genetic testing. To our best knowledge, three studies have investigated genetic diversity and population structures among hemp and marijuana. Sawler et al. (2015) assessed the genetic patterns of 81 marijuana and 43 hemp samples using 14,031 SNPs characterized by genotype-by-sequencing (GBS). Lynch et al. (2016) investigated genetic structure of 340 accessions, which were a mixture of publicly available sequence (WGS and GBS) data and newly sequenced plants, representing three proposed categories based on reported ancestry and/or reported leaf shape: hemp, narrow-leaf drug-type (NLDT, i.e., sativa) and broad-leaf drug type (BLDT, i.e., indica). Dufresnes et al. (2017) conducted genetic analysis of 1324 samples collected from 24 hemp varieties and 15 marijuana strains using 13 SSR markers.

The three studies agreed on statistically significant population differentiation between hemp and marijuana types. However, results were not in agreement concerning whether hemp was more closely related to sativa or indica-types, or the comparison of heterozygosity rates between hemp and marijuana. Sawler et al. (2015) indicated that a hemp population collected from Canada, Europe, and Asia is more genetically related to *C. indica*-type marijuana than to *C. sativa* strains and the hemp population exhibits higher heterozygosity rates than drug-types. Conversely, Lynch et al. (2016) demonstrated that European hemp varieties are more closely related to NLDT than to BLDT, with one exception of a Chinese hemp sample clustering with BLDT and that hemp varieties show less heterozygosity than drug-types, clearly divergent conclusions.

Although the studies differed in their conclusions about hemp's relatedness to indica or sativa groups, both agreed that there is a correlation between genetic structure and reported indica or sativa ancestry using a principal component approach (Sawler et al., 2015) and fastSTRUCTURE (Raj et al., 2014), and FLOCK (Duchesne and Turgeon, 2012) analyses (Lynch et al., 2016) and that these data support a genetic distinction between indica, sativa, and hemp groups. Dufresnes et al. (2017) took a forensic approach and did not attempt to draw a distinction between types of marijuana, simply comparing marijuana with hemp. Their analysis supported that hemp and marijuana are genetically distinct (relating to cannabinoid production, which is generally acknowledged) and that unknown samples could be classified using these markers, but 13 SSR markers in a small population are insufficient to analyze genetic diversity in a way that can ascribe generalizations to the species or understand if there is a genetic basis for distinction outside of THC content.

Both of the studies that utilized substantial genomic coverage used small numbers of hemp samples, 22 in the Lynch et al. (2016) study and 43 in the Sawler et al. (2015) study, which does not fully encompass the genetic diversity of the group. There is also added confusion about distinguishing types of Cannabis because both groups found significant evidence of admixture between all three groups due to natural and human-directed hybridization and reported that marijuana strain names and ancestry data are inherently unreliable. Phylogenetic analyses are helpful in group comparisons, but can be misleading if generalizations are made when species diversity is underrepresented and quality genomes of ancestral species are not available. As more information is added to this debate, a clearer consensus will emerge on true allelic diversity throughout the genome, as well as characterizing population structure. Due to the extensive admixture of these groups and the fact that the basis of their distinction is rooted in a qualitative description of a quantitative phenotype, it is unlikely that population structure will neatly fall into the historically proposed "sativa" and "indica" subgroups.

Germplasm Resources

One of the problems with characterizing genetic diversity and population structure of Cannabis is a lack of access to diverse germplasm. Unlike most crops, no centralized germplasm repository exists for hemp. Americans and Canadians have long histories of producing hemp, but North American germplasm resources were destroyed during Cannabis prohibition. Specifically, a coordinated effort was made to remove Cannabis accessions from gene banks in both the U.S. and Canada around 1980 (Small and Marcus, 2003). Even in countries where hemp was not prohibited, many accessions were lost during periods of political turmoil or through displacement by other crops (Grigoryev, 2017; MultiHemp, 2017). There are a small number of gene banks that store Cannabis germplasm and a few working collections, but all of these organizations act independently and there has not been a collaborative effort of any kind to preserve the Cannabis gene pool.

The Vavilov Institute in Russia maintains the largest collection of approximately 500 accessions of hemp, representing many fiber and seed varieties as well as Chinese landraces (Ranalli, 2004). These are available for research and breeding, but a lack of funding has made maintenance of these accessions difficult for the Institute (Clarke, 1998). Another major gene bank has recently started preserving Cannabis germplasm as well. The Institute of Plant Genetics and Crop Plant Research Gatersleben (IPK) in Germany has a small collection of hemp accessions that are available for research and preservation. This collection contains approximately 55 accessions of cultivated and wild hemp (Graner, 2017). There are also a handful of gene banks that preserve limited collections of mostly local hemp accessions in Hungary, Turkey, Japan, and Italy (Ranalli, 2004). The largest of these collections is in Hungary where 70 local accessions are held, but the others have less than 20 accessions each (Ranalli, 2004).

In addition to gene bank preservation there are some working collections of Cannabis germplasm, but these are not freely available to the public and are not intended for long-term preservation (Ranalli, 2004). The most notable of these is the Dutch Center for Plant Breeding and Reproduction Research (CPRO)/Private Plant Research International (PRI) collection for the Dutch 'National Hemp Program'. It contains 204 accessions comprised of 74 cultivars, 51 landraces, 17 feral samples, and 65 accessions of unknown classification (Bas et al., 2015).

A generous estimate of extant Cannabis accessions would be around 1000 samples total, and it would nearly impossible to access all of these. Additionally, not all of these accessions qualify as (or are) hemp which makes access or possession legally problematic. In comparison, a single germplasm bank at the International Maize and Wheat Improvement Center (CIMMYT) in Mexico maintains approximately 150,000 accessions of wheat (*Triticum aestivum* L.), which are publicly available for research and breeding (Pixley, 2017). Although a direct comparison with a major staple food crop is perhaps unfair, the sheer magnitude of the difference in germplasm resources highlights the fact that Cannabis researchers, breeders, and even farmers face significant challenges in obtaining or creating locally adapted germplasm. It is imperative that a collaborative international effort is undertaken in the near future to preserve the genetic diversity of this potentially important crop before more genes disappear permanently. For more on Cannabis gene bank accessions and the need for coordinated efforts to preserve the Cannabis gene pool, refer to Clarke and Merlin (2016).

Future Directions for Cannabis Genetics Research

One of the primary obstacles to advancing functional genetics research in Cannabis is the lack of an anchored and annotated genome. The highest quality draft genome for Cannabis was published by van Bakel et al. (2011) for PK. Assembly of this genome is difficult because of the lack of a linkage map and the fact that no closely related species have assembled genomes. Even with a genome for a related model species available, recent gene duplications and translocations reduce the accuracy of syntenic alignment (Wicker et al., 2011) and limits assembly to gene coding regions due to rapid evolution of repetitive sequences in non-coding regions (Brunner et al., 2005). It has been recommended that for proper genome assembly at least one segregating population should be sequenced using a whole-genome shotgun sequencing approach to properly align sequences and correct for common errors (Gao et al., 2013; Mascher and Stein, 2014). Although the van Bakel et al. (2011) genome is the only publicly available reference genome, researchers are currently working on implementing structured populations and other approaches in industrial hemp to promote further exploration into gene function validation and genomic studies.

An important goal for breeding high-performance hybrid hemp is properly characterizing heterotic pools within the species. Although a number of genetic diversity studies have been published on Cannabis (Sawler et al., 2015; Lynch et al., 2016; Dufresnes et al., 2017), the limited access to representative samples has not allowed for a full characterization of the germplasm pool, by molecular marker analysis or otherwise. In the United States that problem is exacerbated by the fact that little historical data on hemp performance exists and most currently available germplasm is imported rather than developed locally. This lack of information can be viewed as an opportunity to characterize and curate representative Cannabis populations in a highly documented and organized fashion. Part of this organization should be the development of heterosis breeding. Determining relatedness using molecular markers, along with measuring mid-parent heterosis and assessing parent and F1 performance in mega-environments, has successfully improved classification of maize varieties into heterotic pools (Livini et al., 1992; Reif et al., 2010). Similar strategies could be applied to hemp. As more trial data is collected and next-generation sequencing becomes standard, a comprehensive approach to forming heterotic pools can take place. This will put hemp breeding at an advantage to other crops that developed heterotic pools before these modern tools existed. Although it will still be necessary to develop high-performing inbred lines and regularly test combining ability to assess true heterotic potential, access to modern tools allows breeders to make more rapid and informed choices that can both preserve genetic diversity in the species and maximize heterotic breeding efficiency.

In addition to characterization of agronomic performance, a deeper understanding of genotype by environment interactions (GEI or GxE) should be pursued. Hemp is very sensitive to environmental conditions and its inherent plasticity leads to different phenotypes when soil moisture status, temperatures, or daylength change (Salentijn et al., 2015). This is particularly important regarding cannabinoid content since under the current regulatory framework, production of higher levels of THC can leave a farmer with a crop that must be destroyed rather than a marketable, hemp-based commodity. Using crop modeling to predict the effect of environment on cultivar performance has been successful (Amaducci et al., 2008) but is not a replacement for multi-environment trials (MET). Whether or not research on GEI and MET for hemp occurs in the private or public sector will depend on funding for these types of studies, but a push toward public research in this area would help to advance the understanding of GEI in hemp and hasten the development of locally adapted cultivars.

As was previously mentioned, our understanding of the genetics of important agronomic traits is woefully inadequate in Cannabis. Some initial studies utilizing QTL, GWAS, and TILLING have been performed (Weiblen et al., 2015; Faux et al., 2016; Salentijn et al., 2015; Bielecka et al., 2014), but these are merely first steps in truly understanding genotype to phenotype relationships. More of these types of studies as well as other functional genetic approaches should be used to further our understanding of the control of major traits, including: development of transgenic knockout lines, near-isogenic lines, and comparative genomics with model, crop, and closely related species. Only when the biochemical pathways and genetic architecture of quantitative traits are understood will we be able to fully customize and utilize industrial hemp.

Along with traditional approaches to functional genetics, modern tools may be implemented to advance hemp breeding without identifying causal genes. Marker-assisted selection has become common in many crops, but is primarily only useful for qualitative traits and is limited to QTL that have been verified in breeding populations (Heffner et al., 2009). Genomic selection is a "black box" method that bypasses functional genetics and uses genotype and phenotype data from a training population to predict breeding values and performance of subsequent offspring (Bernardo and Yu, 2007). This approach is able to utilize all genomic information in a way that captures both major and minor allele effects and can more rapidly improve quantitative traits (Chakradhar et al., 2017). It is important to carefully design training populations to mitigate effects of population structure and composition, but genomic selection has been a qualified success in maize breeding programs (Chakradhar et al., 2017). Since hemp faces many similar breeding challenges to maize and high-quality sequence data continues to become more affordable, genomic selection has excellent potential as a breeding method to improve complex traits in hemp.

Summary

The future of research efforts with industrial hemp and Cannabis in general is promising. Although thirty to fifty years ago we saw a massive worldwide effort to eradicate both hemp and marijuana, and its legal status is still variable from place to place today, there has been recent and high-level acknowledgment and acceptance of the medical and industrial uses of the Cannabis plant. New molecular tools are allowing us to look into how the plant functions as well as delve into the origins of the species and classify the wide breadth of diversity observed in the species. Although it is still too early to say whether or not describing official subspecies is warranted, initial studies have supported a genetic basis to distinguishing major gene pools of hemp and marijuana, as well as indica and sativa heritages. Genomics research in Cannabis is still in its infancy, but access to new technologies, combined with less restrictive rules governing industrial hemp research, promises a wealth of information to come. More collaborative, government-funded research, like the European Union MultiHemp project, is an absolute necessity to advance hemp research in a rigorous way that contributes to the evolution of the nascent industry. The multitude of uses possible for Cannabis warrants a methodological approach to fully understanding and characterizing the genetic architecture of important traits, so that the plant can be optimized for a variety of tasks. Only by modernizing our approach to understanding Cannabis genetics and genomics will it be possible to utilize and regulate production of this plant in a way that is truly beneficial, and in the most efficient manner possible. Every other crop of significant economic importance has been characterized in this way. We propose that it is time for Cannabis research to catch up, so that impactful, plant-based solutions are not overlooked or under-utilized.

Literature Cited

Amaducci, S., M. Colauzzi, G. Bellocchi, and G. Venturi. 2008. Modelling post-emergent hemp phenology (Cannabis sativa L.): Theory and evaluation. Eur. J. Agron. 28(2):90–102. doi:10.1016/j.eja.2007.05.006

Andre, C.M., J.-F. Hausman, and G. Guerriero. 2016. Cannabis sativa: The plant of the thousand and one molecules. Front. Plant Sci. 7:9.

van Bakel, H., J.M. Stout, A.G. Cote, C.M. Tallon, A.G. Sharpe, T.R. Hughes, and J.E. Page. 2011. The draft genome and transcriptome of Cannabis sativa. Genome biology 12(10): R102. doi:10.1186/gb-2011-12-10-r102

Bas, N., M. Toonen, and L. Trindade. 2015. Current status of the Dutch Hemp collection. Korea 3(1):4.

Bernardo, R. Breeding for quantitative traits in plants. No. 576.5 B523. Stemma Press, 2002.

Bernardo, R., and J. Yu. 2007. Prospects for genomewide selection for quantitative traits in maize. Crop Sci. 47(3):1082–1090. doi:10.2135/cropsci2006.11.0690

Bielecka, M., F. Kaminski, I. Adams, H. Poulson, R. Sloan, Y. Li, T.R. Larson, T. Winzer, and I.A. Graham. 2014. Targeted mutation of D12 and D15 desaturase genes in hemp produce major alterations in seed fatty acid composition including a high oleic hemp oil. Plant Biotechnol. J. 12(5):613–623. doi:10.1111/pbi.12167

Bócsa, I. 1999. Genetic improvement: Conventional approaches. Advances in Hemp Research, Haworth Press, Binghamton, NY. p. 153-184.

Booth, Judith K., Jonathan E. Page, and Jörg Bohlmann. 2017. Terpene synthases from Cannabis sativa. PloS one 12(3): E0173911. doi:10.1371/journal.pone.0173911

Brunner, S., K. Fengler, M. Morgante, S. Tingey, and A. Rafalski. 2005. Evolution of DNA sequence nonhomologies among maize inbreds. Plant Cell 17(2):343–360. doi:10.1105/tpc.104.025627

Chakradhar, T., V. Hindu, and P.S. Reddy. 2017. Genomic-based-breeding tools for tropical maize improvement. Genetica (The Hague) 145(6):1–15.

Clarke, Robert C. 1998. Maintenance of Cannabis germplasm in the Vavilov Research Institute gene bank-five year report. International Hemp Association, Amsterdam, The Netherlands.

Clarke, Robert C., and Mark D. Merlin. 2016. Cannabis domestication, breeding history, present-day genetic diversity, and future prospects. Critical Reviews in Plant Sciences 35(5-6): 293–327. doi:10.1080/07352689.2016.1267498

Devinsky, O., M.R. Cilio, H. Cross, J. Fernandez-Ruiz, J. French, C. Hill, R. Katz, V. Di Marzo, D. Jutras-Asward, W.G. Notcutt, J. Martinez-Orgado, P.J. Robson, B.G. Rohrback, E. Thiele, B. Whalley, and D. Friedman. 2014. Cannabidiol: Pharmacology and potential therapeutic role in epilepsy and other neuropsychiatric disorders. Epilepsia 55(6):791–802. doi:10.1111/epi.12631

Dewey, L.H. 1927. Hemp varieties of improved type are result of selection. In: USDA, editor, Year-book of the USDA. USDA, Washington, D.C.

Divashuk, Mikhail G., O.S. Alexandrov, O.V. Razumova, I.V. Kirov, and G.I. Karlov. 2014. Molecular cytogenetic characterization of the dioecious Cannabis sativa with an XY chromosome sex determination system. PLoS One 9(1): E85118.

Duchesne, P., and J. Turgeon. 2012. FLOCK provides reliable solutions to the "number of populations" problem. J. Hered. 103(5):734–743. doi:10.1093/jhered/ess038

Dufresnes, C., C. Jan, F. Bienert, J. Goudet, and L. Fumagalli. 2017. Broad-scale genetic diversity of Cannabis for forensic applications. PloS one 12(1): E0170522. doi:10.1371/journal.pone.0170522

Faux, A.-M., A. Berlin, N. Dauguet, and P. Bertin. 2014. Sex chromosomes and quantitative sex expression in monoecious hemp (Cannabis sativa L.). Euphytica 196(2):183–197. doi:10.1007/s10681-013-1023-y

Faux, A.-M., X. Draye, M.-C. Flamand, A. Occre, and P. Bertin. 2016. Identification of QTLs for sex expression in dioecious and monoecious hemp (Cannabis sativa L.). Euphytica 209(2):357–376. doi:10.1007/s10681-016-1641-2

Felberbaum, M., and S. Walsh. 2018. FDA approves first drug comprised of an active ingredient derived from marijuana to treat rare, severe forms of epilepsy. FDA News Release. FDA, Washington, D.C. https://www.fda.gov/newsevents/newsroom/pressannouncements/ucm611046.htm (Accessed 13 Mar. 2019).

Gao, Z.-Y., S.-C. Zhao, W.-M. He, L.-B. Guo, Y.-L. Peng. J-J. Wang, X.-S. Guo. et al. 2013. Dissecting yield-associated loci in super hybrid rice by resequencing recombinant inbred lines and improving parental genome sequences. Proc. Natl. Acad. Sci. USA 110(35):14492–14497. doi:10.1073/pnas.1306579110

Graner, A. Genebank department. IPK Gatersleben, Berlin, Germany. http://www.ipk-gatersleben.de/en/genebank/ (Accessed 18 Sept. 2017).

Grigoryev, S. Hemp (Cannabis sativa L.) genetic resources at the VIR: From the collection of seeds, through the collection of sources, towards the collection of donors of traits. Vavilov Institute, St. Petersburg, Russia. http://www.vir.nw.ru/hemp/hemp1.htm (Accessed 18 Sept. 2017).

Grotenhermen, Franjo, and Kirsten Müller-Vahl. 2016. Medicinal uses of marijuana and cannabinoids. Critical Reviews in Plant Sciences 35(5-6): 378-405. doi:10.1080/07352689.2016.1265360

Haunold, A. 1991. Cytology and cytogenetics of hops. Dev. Plant Genet. Breed. 2:551–563.

Heffner, E.L., M.E. Sorrells, and J.-L. Jannink. 2009. Genomic selection for crop improvement. Crop Sci. 49(1):1–12. doi:10.2135/cropsci2008.08.0512

Hennink, Sebastiaan. 1994. Optimisation of breeding for agronomic traits in fibre hemp (Cannabis sativa L.) by study of parent-offspring relationships. Euphytica 78(1–2): 69–76.

Hillig, K.W., and P.G. Mahlberg. 2004. A chemotaxonomic analysis of cannabinoid variation in Cannabis (Cannabaceae). Am. J. Bot. 91(6):966–975. doi:10.3732/ajb.91.6.966

Hillig, K.W. 2005. Genetic evidence for speciation in Cannabis (Cannabaceae). Genet. Resour. Crop Evol. 52(2):161–180. doi:10.1007/s10722-003-4452-y

Horkay, E. 1982. Primary and secondary fibre-cell ratio as affected by selection for increasing fibre content. (In Hungarian). Novenytermeles 31:297–301.

Karlov, G.I., T.V. Danilova, C. Horlemann, and G. Weber. 2003. Molecular cytogenetics in hop (Humulus lupulus L.) and identification of sex chromosomes by DAPI-banding. Euphytica 132(2):185–190. doi:10.1023/A:1024646818324

Kim, C., D. Zhang, S.A. Auckland, L.K. Rainville, K. Jakob, B. Kronmiller, E.J. Sacks, M. Deuter, and A.H. Paterson. 2012. SSR-based genetic maps of Miscanthus sinensis and M. sacchariflorus, and their comparison to sorghum. Theor. Appl. Genet. 124(7):1325–1338. doi:10.1007/s00122-012-1790-1

Laursen, L. 2015. Botany: The cultivation of weed. Nature 525(7570):S4–S5. doi:10.1038/525S4a

Li, H.-L. 1973. An archaeological and historical account of Cannabis in China. Econ. Bot. 28(4):437–448. doi:10.1007/BF02862859

Livini, C., P. Ajmone-Marsan, A.E. Melchinger, M.M. Messmer, and M. Motto. 1992. Genetic diversity of maize inbred lines within and among heterotic groups revealed by RFLPs. Theor. Appl. Genet. 84(1–2):17–25. doi:10.1007/BF00223976

Lynch, R.C., D. Vergara, S. Tittes, K. White, C.J. Schwartz, M.J. Gibbs, T.C. Ruthenburg, K. deCesare, D.P. Land, and N.C. Kane. 2016. Genomic and chemical diversity in Cannabis. Crit. Rev. Plant Sci. 35(5–6): 349-363. doi:10.1080/07352689.2016.1265363

Mascher, M., and N. Stein. 2014. Genetic anchoring of whole-genome shotgun assemblies. Front. Genet. 5:208.

Mechoulam, R., A. Shani, B. Yagnitinsky, Z. Ben-Zvi, P. Braun, and Y. Gaoni. 1970. Some aspects of cannabinoid chemistry. In: C.R.B. Joyce and S.H. Curry, editors, The botany and chemistry of Cannabis. Churchill, London, UK. p. 93–115.

de Meijer, E.P.M., M. Bagatta, A. Carboni, P. Crucitti, V.M.C. Moliterni, P. Ranalli, and G. Mandolino. 2003. The inheritance of chemical phenotype in Cannabis sativa L. Genetics 163(1):335–346.

de Meijer, E.P.M., K.M. Hammond, and A. Sutton. 2009. The inheritance of chemical phenotype in Cannabis sativa L.(IV): Cannabinoid-free plants. Euphytica 168(1):95–112. doi:10.1007/s10681-009-9894-7

Ming, R., A. Bendahmane, and S.S. Renner. 2011. Sex chromosomes in land plants. Annu. Rev. Plant Biol. 62:485–514. doi:10.1146/annurev-arplant-042110-103914

MultiHemp. 2017. MultiHemp. Università Cattolica del Sacro Cuore, Piacenza, Italy. http://multihemp.eu/project/. [2017 is year accessed].

Natsume, S., H. Takagi, A. Shiraishi, J. Murata, H. Toyonaga, J. Patzak, M. Takagi, H. Yaegashi, A. Uemura, C. Mitsuoka, K. Yoshida, K. Krofta, H. Satake, R. Terauchi, E. Ono. 2014. The draft genome of hop (Humulus lupulus), an essence for brewing. Plant Cell Physiol. 56(3):428–441. doi:10.1093/pcp/pcu169

Neve, R.A. 1958. Sex chromosomes in the hop Humulus lupulus. Nature 181(4615):1084–1085. doi:10.1038/1811084b0

Paterson, A.H. 1995. Molecular dissection of quantitative traits: Progress and prospects. Genome Res. 5(4):321–333. doi:10.1101/gr.5.4.321

Pixley, K. Germplasm research. CIMMYT, El Batan, Mexico. http://www.cimmyt.org/germplasm-bank/ (Accessed 18 Sept. 2017).

Raj, A., M. Stephens, and J.K. Pritchard. 2014. fastSTRUCTURE: Variational inference of population structure in large SNP data sets. Genetics 197(2):573–589. doi:10.1534/genetics.114.164350

Mohan Ram, H.Y., and R. Sett. 1982. Induction of fertile male flowers in genetically female Cannabis sativa plants by silver nitrate and silver thiosulphate anionic complex. Theor. Appl. Genet. 62(4):369–375. doi:10.1007/BF00275107

Ranalli, P. 2004. Current status and future scenarios of hemp breeding. Euphytica 140(1–2):121–131. doi:10.1007/s10681-004-4760-0

Razumova, O.V., O.S. Alexandrov, M.G. Divashuk, T.I. Sukhorada, and G.I. Karlov. 2016. Molecular cytogenetic analysis of monoecious hemp (Cannabis sativa L.) cultivars reveals its karyotype variations and sex chromosomes constitution. Protoplasma 253(3):895–901. doi:10.1007/s00709-015-0851-0

Reif, J.C., S. Fischer, T.A. Schrag, K.R. Lamkey, D. Klein, B.V. Dhillon, H.F. Utz, and A.E. Melchinger. 2010. Broadening the genetic base of European maize heterotic pools with U.S. Cornbelt germplasm using field and molecular marker data. Theor. Appl. Genet. 120(2):301–310. doi:10.1007/s00122-009-1055-9

Russo, E.B. 2011. Taming THC: Potential Cannabis synergy and phytocannabinoid terpenoid entourage effects. Br. J. Pharmacol. 163(7):1344–1364.

Sakamoto, K., Y. Akiyama, K. Fukui, H. Kamada, and S. Satoh. 1998. Characterization; genome sizes and morphology of sex chromosomes in hemp (Cannabis sativa L.). Cytologia (Tokyo) 63(4):459–464. doi:10.1508/cytologia.63.459

Salentijn, E.M.J., Q. Zhang, S. Amaducci, M. Yang, and L.M. Trinidade. 2015. New developments in fiber hemp (Cannabis sativa L.) breeding. Ind. Crops Prod. 68: 32–41.

Sawler, J., J.M. Stout, K.M. Gardner, D. Hudson, J. Vidmar, L. Butler, J.E. Page, and S. Myles. 2015. The genetic structure of marijuana and hemp. PloS one 10(8): E0133292. doi:10.1371/journal.pone.0133292

Shephard, H.L., J.S. Parker, P. Darby, and C.C. Ainsworth. 2000. Sexual development and sex chromosomes in hop. New Phytol. 148(3):397–411. doi:10.1046/j.1469-8137.2000.00771.x

Shoyama, Y., M. Yagi, I. Nishioka, and T. Yamauchi. 1975. Biosynthesis of cannabinoid acids. Phytochemistry 14:2189–2192. doi:10.1016/S0031-9422(00)91096-3

Sinotô, Y. 1929. On the tetrapartite chromosome in Humulus lupulus. Proc. Imp. Acad. (Tokyo)5(1):46–47. doi:10.2183/pjab1912.5.46

Small, E., and A. Cronquist. 1976. A practical and natural taxonomy for Cannabis. Taxon 25(4):405–435. doi:10.2307/1220524

Small, E., and D. Marcus. 2003. Tetrahydrocannabinol levels in hemp (Cannabis sativa) germplasm resources. Econ. Bot. 57(4):545–558. doi:10.1663/0013-0001(2003)057[0545:TLIHCS]2.0.CO;2

Stemple, Derek L. 2013. So, you want to sequence a genome... Genome biology 14(7):128. doi:10.1186/gb-2013-14-7-128

Taura, F., S. Morimoto, . 1995. First direct evidence for the mechanism of. DELTA. 1-tetrahydrocannabinolic acid biosynthesis. J. Am. Chem. Soc. 117(38):9766–9767. doi:10.1021/ja00143a024

Till, B.J., T. Zerr, L. Comai, and S. Henikoff. 2006. A protocol for TILLING and ecotilling in plants and animals. Nature protocols 1(5): 2465–2477. doi:10.1038/nprot.2006.329

van den Broeck, H.C., C. Maliepaard, M.J.M. Ebskamp, M.A.J. Toonen, and A.J. Koops. 2008. Differential expression of genes involved in C 1 metabolism and lignin biosynthesis in wooden core and bast tissues of fibre hemp (*Cannabis sativa* L.). Plant Sci. 174(2):205–220. doi:10.1016/j.plantsci.2007.11.008

Vergara, D., H. Baker, K. Clancy, K.G. Keepers, P. Mendieta, C.S. Pauli, S.B. Tittes, K.H. White, and N.C. Kane. 2016. Genetic and genomic tools for Cannabis sativa. Critical Reviews in Plant Sciences 35(5–6): 364–377. doi:10.1080/07352689.2016.1267496

Watts, Geoff. 2006. Science commentary: Cannabis confusions. BMJ: British Medical Journal 332(7534): 175–176.

Weiblen, G.D., J.P. Wenger, K.J. Craft, M.A. ElSohly, Z. Mehmedic, E.L. Treiber, and M.D. Marks. 2015. Gene duplication and divergence affecting drug content in Cannabis sativa. New Phytol. 208(4):1241–1250.

Wicker, T., K.F.X Mayer, H. Gundlach, M. Martis, B. Steuernagel, U. Scholz, H. Šimková, M. Kubaláková, et al. 2011. Frequent gene movement and pseudogene evolution is common to the large and complex genomes of wheat, barley, and their relatives. The Plant Cell Online 23(5):1706–1718. doi:10.1105/tpc.111.086629

Yang, M.-Q., R. van Velzen, F.T. Bakker, A. Sattarian, D.-Z. Li, and T.-S. Yi. 2013. Molecular phylogenetics and character evolution of Cannabaceae. Taxon 62(3):473–485. doi:10.12705/623.9

Zhang, D., N.J. Pitra, M.C. Coles, E.S. Buckler, and P.D. Matthews. 2016. Non-Mendelian inheritance of SNP markers reveals extensive chromosomal translocations in dioecious hops (*Humulus lupulus* L.). bioRxiv 069849.

Chapter 7: Economic Issues and Perspectives for Industrial Hemp

Tyler B. Mark* and Will Snell

Introduction

The "economics" of hemp is a complex topic. Although the crop has been produced and marketed for thousands of years, the current hemp industry presently contains many economic uncertainties for hemp producers, processors, manufacturers, retailers, input suppliers, and consumers. While many unknowns surround the economics of hemp, three definitive statements can be made about the evolving hemp industry:

- Hemp can be used as an input for thousands of products.

- Sales of hemp products in the United States and worldwide currently represent a relatively small market share of overall food, textile, personal care products, pharmaceutical or nutraceutical products, and sales from other sectors, but have been growing at a relatively brisk pace in recent years.

- Global production of hemp has declined considerably since the 1950s, but has been rebounding over the past decade in response to growing consumer demand for hemp products, policy changes, infrastructure and business investment, and improved production practices.

This environment is creating much enthusiasm and perceived opportunities for the crop, but it is not without its economic challenges and policy and regulatory uncertainties. Some of these uncertainties have been be addressed by the enactment of the 2018 Farm Bill, which included language to remove industrial hemp from the controlled substance list, allows for hemp farmers to be eligible for federal crop insurance and allows hemp researchers to apply for competitive federal grants. Despite legislative approval, regulatory uncertainties among federal agencies such the FDA and DEA still exist.

Ultimately, business models for profit-maximizing firms contemplating using hemp in their products must find hemp cost competitive with other competing inputs such as synthetic or other natural fibers, alternative oils, and other health supplements and therapeutic compounds. For farmers, hemp must be profitable relative to other potential crops and agricultural enterprises and competitive with hemp imported from competing countries. Consumer demand for hemp

T.B. Mark, University of Kentucky, Lexington, KY; W. Snell, University of Kentucky Extension, Lexington, KY. *Corresponding author (tyler.mark@uky.edu)

doi:10.2134/industrialhemp.c7
Industrial Hemp as a Modern Commodity Crop. D.W. Williams, editor.

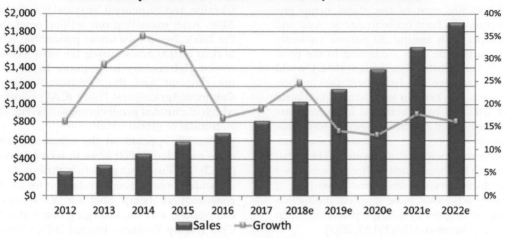

U.S. Hemp-Based Products Sales, 2012-2022e

Source: *Hemp Business Journal* estimates ($ mil., consumer sales)

Fig. 1. U.S. Hemp-Based Product Sales, 2012 through 2020.

will be shaped by the utility they receive from purchasing hemp products, which includes perceived health and environmental benefits, subject to price levels for hemp products and income constraints.

Economists are challenged in evaluating this crop's economic potential given multiple uses for this crop from different parts of the plant, a highly variable and unpredictable policy and regulatory environments across U.S. states, and limited market and farm-level data. This chapter will provide a historical review of global hemp production and products markets, identify specific economic issues facing producers, and present some farm-level economic analysis.

U.S. Hemp Product Market

The Hemp Industries Association (HIA), a non-profit trade association representing U.S. hemp businesses, estimates that U.S. retail sales of hemp products totaled at least $820 million in 2017, shown in Fig. 1.

This includes an increase of 16% from 2016 levels and a continuation of double-digit percentage growth since 2012. According to HIA, the rapidly growing CBD product market now comprises the largest share of hemp product sales (23%), followed by personal care products (22%), food (17%), industrial applications such as car parts

(18%), and textiles (13%). The Hemp Business Journal (2018) projects that the hemp industry will grow to $1.9 billion in sales by 2022, led by CBD-based products and industrialized applications.

The growth in hemp product demand, along with hemp production restrictions before the passage of the 2014 Farm Bill, boosted hemp-related imports into the United States. According to various trade data accumulated by the Congressional Research Service (CRS), U.S. hemp product imports exceeded $78 million in 2015, almost doubling from its 2013 and 2014 levels. However, U.S. hemp imports have been declining in recent years given production in the United States, down to $69 million in 2016 and $67 million in 2017. (Johnson, 2018). It is important to note that a limitation to this data is that it does not include finished products, such as hemp-based clothing or other products including construction materials, carpets, or hemp-based paper products, which underestimates total imports of hemp products into the United States. Most of the imports are classified as hemp seeds, used primarily for hemp-based foods, supplements, and body care products. Canada is currently the primary supplier of hemp seed to the United States and China is the largest supplier of raw and processed hemp fiber (FAOSTAT, 2017).

Global Hemp Production

Historically most of the global hemp production has been concentrated in Asia and Europe, with around 30 nations currently producing the crop. According to the Food and Agriculture Organization (FAO), world hemp acreage fell from over two million acres in the 1960s to less than 150,000 acres by 2010, but has rebounded in recent years on the heels of expanding demand for a wide variety of consumer products and the establishment of new production areas. The global growth has been much higher for hemp seed production versus fiber production (Fig. 2). Improved yields have enabled production volume to grow by a larger percentage than acreage harvested (FAOSTAT, 2017).

China has historically been a major producer and exporter of hemp as inexpensive labor, favorable government policy, and processing infrastructure led to a dominant global market share during the twentieth century. However, according to the FAO database, Chinese hemp production has declined considerably over the past several decades, with area harvested totaling around 30,000 acres in recent years. Despite its production decline, China remains a global leader in supplying low-cost fiber for textile hemp products.

Europe has also been a significant player in the global hemp market, with France being the dominant European hemp producing nation, along with the Netherlands, Lithuania, Italy, Russia, Romania, Ukraine, and Hungary (Carus and Sarmento 2016). After exceeding one million acres in the 1960s, hemp area under production fell to near 50,000 acres during the early years of the 21st century in response to declining demand and the European Union's Common Agricultural Policy (CAP) reform which eliminated producer subsidies (Fig. 3).

Similar to global trends, European hemp production appears on the rebound with acres exceeded 70,000 acres in recent years (Fig. 3). The European growth is attributed to expanded organic seed production for food consumption, hemp fiber for automobile composites, and the emerging demand for CBD (Hemp Business Journal 2017a; European Industrial Hemp Association 2017)

While hemp production has been prevalent in several European and Asian nations for centuries, Canada's hemp industry is a relative newcomer and has been evolving over the past two decades. The Canadian government initially issued research licenses in 1994 to grow industrial hemp on an experimental basis. Beginning in 1998, commercial production became legal with licenses and other regulatory services provided by the Office of Controlled Substances of Health Canada. During its early years of development, Canada's hemp industry experienced a lot of volatility and some challenging times in response to speculative

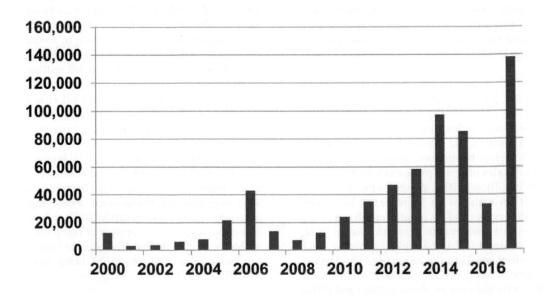

Fig. 2. World Hemp Seed and Fiber Acres. Source: (FAOSTAT, 2017).

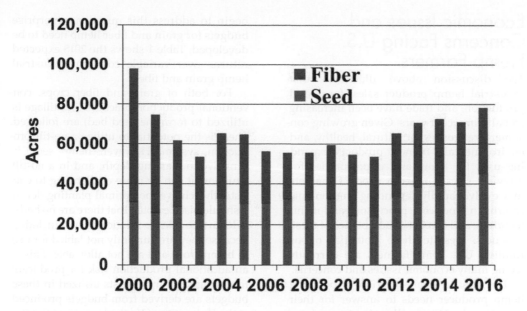

Fig. 3. European Hemp Seed and Fiber Acres, 2000–2014. Source: (FAOSTAT 2017).

demand, competition from other crops, and changes in U.S. import rules. Canadian hemp acreage (which is excluded from the FAO database presented above) grew steadily from 2008 to 2014, increasing to around 100,000 acres in 2014 as improved processing technology, infrastructure development, research, government financial support, product development, and growing import demand by the United States boosted sales (Fig. 4).

Canadian hemp planted area retreated to around 85,000 acres in 2015 and slipped to about half of that in 2016 in response to excessive inventories. However, acreage rebounded in 2017 to a record 138,000 acres with anticipation of increased export sales to South Korea and the United States. As export demand declined, Canadian hemp acres and prices reportedly declined sharply in 2018. (AgCanada, 2018). Canadian hemp production is primarily grain production as the Canadian market has struggled with finding profitable markets for fiber (Amason, 2017; Syngenta Canada, 2017). However, legislative changes in 2018 now allow Canadian hemp farmers to produce for the growing CBD market. (Hemp Industry Daily, 2018)

Hemp Production in the United States

The passage of the 2014 U.S. Farm Bill provided a legal framework through a highly regulated research platform administered through state Departments of Agriculture and agricultural universities. Following passage and adoption of supporting state-level legislation, these institutions are charged with administering and managing the reintroduction of production and marketing of industrial hemp in the United States. As of 2018, forty U.S. states have adopted various legislative measures to address the production of hemp in their respective states, with several others considering legislation (Vote Hemp, 2017). The USDA does not track production data related to hemp. Analysts must rely on unofficial state and industry data. Also, one should note there are significant annual discrepancies between the reported approved registered acres, planted acres, and harvested acres in nearly every state participating in the pilot research program. Approved U.S. hemp area totaled around 2000 acres in 2014, with less than 200 acres reportedly planted and harvested (primary in Colorado and Kentucky) due to challenges related to access to seed, germination issues, and overall knowledge on production practices (Fig. 5).

Economic Issues and Concerns Facing U.S. Hemp Farmers

The discussion above illustrates that industrial hemp product sales, farm-level production, and trade have been increasing globally in recent years. Given growing consumer preferences for natural, healthy, and environmentally friendly products, expanding uses for hemp and opportunities for growing sales for hemp product markets may evolve rapidly depending on economic returns to producers, processors, and manufacturers, and consumer acceptance of these products. Despite these potential opportunities, U.S. hemp farmers are currently facing many economic issues and concerns.

The first critical question a potential hemp producer needs to answer for their operation is: **How will net returns from industrial hemp being inserted into the crop rotation impact the short and long-term profitability of the operation?** To begin to address this question, enterprise budgets for grain and fiber hemp need to be developed. Table 1 shows the 2018 expected returns over variable costs for industrial hemp grain and fiber.

For both of grain and fiber crops, conventional production practices (i.e., tillage is utilized to form the seed bed) are followed. There is the potential to utilize a no-till production system. However, hemp seed is sensitive to planting depth, and in a no-till situation, it can be more challenging to consistently achieve the optimal planting depth. It should also be noted that there are no herbicide, insecticide, or fungicide costs included because they are currently not labled for use in hemp crops and are not allowable. This is an additional production risk for producers to manage. All input costs utilized in these budgets are derived from budgets produced at the University of Kentucky or from the custom harvest survey (Halich 2018a, b). Hemp grain and fiber prices are representative of 2018 pricing in the Kentucky market. Yields

Table 1. 2018 Expected returns above variable costs (RAVC) for Kentucky industrial grain and fiber hemp.

	Hemp grain				Hemp fiber			
	Quant.	Unit	Price	Total	Quant.	Unit	Price	Total
Gross returns per acre								
Hemp grain	768	lb	$0.70	$537.51	6,400	lb	$0.08	$502.43
Total revenue				$537.51				$502.43
Variable costs per acre								
Seed	30	lb	$4.80	$144.02	50	lb	$2.30	$114.96
Nitrogen (solid urea 46% N)	100	units	$0.40	$40.00	50	units	$0.40	$20.00
Phosphorus (P_2O_5)	30	units	$0.36	$10.80	45	units	$0.36	$16.20
Potassium (K_2O)	45	units	$0.30	$13.50	35	units	$0.30	$10.50
Lime - Delivered and spread	0.3	ton	$20.00	$6.00	0.3	ton	$20.00	$6.00
Disk Harrow	1	unit	$10.50	$10.50	1	unit	$10.50	$10.50
Grain Drill	1	10 feet	$17.50					
Harvest Cost	1	25 feet rigid	$30.00	$30.00				$0.00
Haul Hemp Grain	1		$94.24	$94.24				$0.00
Custom work	1	acre	$0.00	$0.00	1	acre	$71.50	$71.50
Cash rent	1	acre	$150.00	$150.00	1	acre	$150.00	$150.00
Other variable costs		acre			1	acre	$111.70	$111.70
Interest on operating capital			$7.28					$7.28
Unallocated labor			$7.47					$7.47
Total variable costs per acre			$531.30					$543.61
Return above variable costs per acre		Grain RAVC	$6				Fiber RAVC	-$41.18
Breakeven yield at $0.70 lb⁻¹	759	lb acre⁻¹ to cover variable costs			Breakeven yield at $0.078505288			6925
Breakeven cost at 768 lb		$0.69 lb⁻¹ to cover variable costs			Breakeven cost at 6400 lb			$0.08

for industrial hemp vary widely for a number of different reasons. First, as discussed above, there are no chemicals that can be utilized in the production of hemp, so to control weeds it is imperative that hemp shade out the other competing weeds. This has been the primary reason cited by Kentucky farmers for the failure of their crop. The passage of the 2018 Farm Bill could change this situation dramatically with hemp being removed from the controlled substance list and agricultural chemical companies begin the research and development process to add hemp to their labels. Second, hemp yields vary widely as a result of seed genetics that were originally adapted to the environments and latitudes of other regions of the globe. It is expected as plant breeding efforts and localized production of certified industrial hemp seed increases then yields will also increase and stabilize. In support of this statement, yields measured from hemp grain and fiber variety trials conducted at the University of Kentucky in 2018 reached record levels; 1706 and 10,560 pounds per acre of dry grain and retted, dry, baled fiber, respectively. These record yields would change the RAVC values as reported in Table 1 above significantly to $662.90 (grain) and $301.19 (fiber), both of which are competitive with current RAVC values for corn and soybeans.

As the industrial hemp market continues to mature, it is expected that the industrial hemp industry will be similar to existing homogenous grain markets, where short-term profits will entice increased supplies that will eventually result in lower prices and ultimately generating a nominal rate of return keeping only the lowest cost producers in the industry. One potential exception to this might be the markets for CBD. Cannabidiol currently offers higher economic returns for producers, but also possess more volatile financial, policy, and regulatory risk than markets for hemp fiber and grain. Likewise, lucrative short-run profits that may exist during the early years of this emerging industry will likely lure additional supply worldwide, which will diminish future profit potential and commoditize the market, unless barriers to entry for this market are created. At this writing, it is unknown if the U.S. federal government will regulate CBD and other cannabinoids as pharmaceutical compounds available only by prescription. If this level of regulation is enforced, production will likely be much smaller in scale so as to increase the potential for high levels of quality control. An additional factor that sets hemp production apart is that it will likely have additional regulatory costs (THC testing, application fees, etc.), input restrictions (e.g., chemicals) and perhaps specialized equipment that is not encountered by other crops. This creates additional costs and management challenges that potential hemp producers will have to deal with as the industry grows.

The second question producers need to consider is "How will they handle price volatility amidst increasing growing producer interest among many U.S. states and global competition potentially leading to oversupply challenges"? Another unique feature that sets this crop apart from our traditional commodities is that there is minimal ability to manage the inherent risks associated with price and/or yield. Crop insurance, farm programs, futures markets, and marketing cooperatives to reduce producer price and income risks are not currently present in the hemp market. Producers need to look no further than our neighbors to the north, Canada, to see how price volatility impacts production as shown in Fig. 4. Over the last 20 years, Canadian producers have had significant swings in their acreages of grain production as the price can vary widely from year to year. Producers in the United States are used to a production environment that provides safety nets if yield or price for their commodity drops too low. This is not currently the case for industrial hemp. Producers need to understand and be willing to lose their investment in the crop if it fails, the processor goes out of business, or the policy environment changes. If these are not risks the producer is willing to accept or does not have the financial ability to absorb, then hemp may not be the right crop for their operation until these conditions are ameliorated or become more stable.

This does not mean that producers should not consider industrial hemp production, but that a higher degree of caution and consideration should be taken before entering the market. Passage of the 2018 Farm Bill does soften some of these financial risks with the producers being eligible for federal crop insurance, which would provide significantly improved structure to risk management plans. Additionally, it might also be advisable to

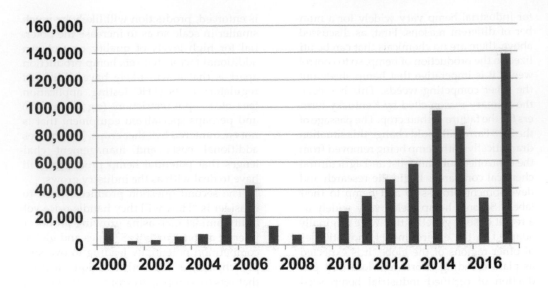

Fig. 4. Canadian Hemp Acres, 2000–2014.

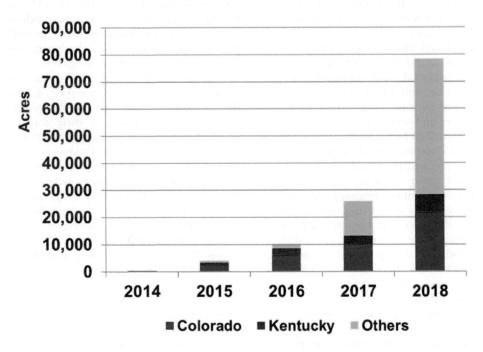

Fig. 5. U.S. Hemp Planted Acres, 2014–2016. Registered U.S. hemp acres increased to nearly 7000 acres in 2015 with around 4000 acres being planted. Industry and state reports indicated that nineteen U.S. states approved more than 25,000 hemp acres in 2017. Kentucky and Colorado in 2017 were joined by Oregon and North Dakota as the dominant hemp-producing states, accounting for over 75% of U.S. acreage in 2017. While approved U.S. hemp acres surpassed 25,000 acres in 2017, production reports reveal that hemp harvested acres were likely around 10,000 acres in 2017 (Vote Hemp, 2017; Hemp Business Journal 2017b). Early indications reveal that 2018 hemp acres in the United States may have more than doubled in 2018, with expectations that passage of the 2018 Farm Bill will entice additional production growth in coming years.

consider alternative market structures, such as vertically-integrated production models to share risk among buyers and sellers and allow buyers greater control over input, use and production practices to control the quantity and quality of a highly-regulated crop.

Beyond just managing risk for their operation, producers also need to consider the impact that production from other countries can have on their market. For the United States to be a viable hemp producer, more research is needed to determine the growing areas in the United States that provide a competitive advantage regarding growing seasons and soils, management to generate higher yields, and better quality hemp. Undoubtedly, competition from Canada and Europe (with established infrastructure, management expertise, and markets), and China (with access to lower wages and lower regulatory standards) will provide fierce competition for U.S. hemp farmers, processors, and associated companies. Additionally, Canada has changed the rules to allow hemp floral material to be harvested and processed for CBD (Nichols, 2017). This could have a significant impact on CBD prices in the United States that are already very volatile and unpredictable. This is the primary reason for limited discussion in this chapter on CBD. There are too many risks in the market even to make an educated guess. Uruguay, the first country in the world to make all forms of cannabis production legal, is another example of a country that is entering the market. They are a small country but have productive soils and the agricultural technology in place to become a significant player in the production of industrial hemp. Additionally, they have already removed many of the regulatory hurdles that we have in the United States. Global competition is going to continue to be an issue that U.S. producers and others active in the industry need to consider.

A crucial third question producers need to consider is: **What is the availability of credit, labor, certified seed, and research for hemp?** Access to inputs at competitive prices will be critical to control costs and enhance yields and quality. Currently, many financial institutions are reluctant or will not provide capital for this crop. Therefore, some producers are utilizing operating capital they have secured for other crops to grow hemp, and with minimal ways to manage risk in hemp production, this can be very dangerous.

Once more is known about the market and prices become more stable, this situation is expected to improve as long as hemp becomes an entirely legal crop to produce.

Labor availability is another consideration for producers. For grain and fiber, it is not currently an issue because these production systems are already mechanized. However, multiple CBD production systems are still under investigation. Cannabidiol production will most likely require significantly more labor for practices like transplant production, transplanting, rogueing male hemp plants to prevent pollination, and for the harvest of flower material. Access to labor could become an issue for producers given the current and rapidly evolving immigration policies in the United States.

Certified seed from varieties suitable for the specific regions that industrial hemp is being produced is another hurdle. Hemp has been widely produced all across the globe, but genetics suitable for the United States have not been investigated for almost six decades. Therefore, we are utilizing seed from regions of the world that have different latitudes and microclimates, and the countries they are being produced in have different regulatory standards for seed quality relative to the United States. Domestic seed production for hemp seed is just now beginning in the United States, and it will take time for the market to catch up with the demand for seed. However, once it does, seed costs should decrease for producers and yields should rise. This will improve hemp competitiveness with other, existing crops in a producer's rotation.

One of the drivers of the United States competitiveness in the global grain markets is that producers have access to cutting-edge research and the newest technology. Hemp production in the United States has been nonexistent since World War II. Therefore, research into the crop has been essentially nonexistent. Currently, we are relying on anecdotal evidence and research that has been done in other countries with different resource endowments. The hemp provisions in the 2018 Farm Bill should allow for an increased hemp research effort across the country, but it is going to take years for the research to catch up with the questions needing to be addressed as producers make management decisions for the near-term. To date, essentially all funding for U.S. industrial hemp research has been derived from corporate sources. While this is not inherently negative, sometimes corporate funding is aligned with a need to maintain the information derived from the work internal to the corporation or considered proprietary. While these projects may contribute significantly to the general body of knowledge, they can also reduce the widespread dissemination of information to the general public relative to other, perhaps publicly-funded sources. As of 2018, some state governments are investing in university-based industrial hemp research activities. Additionally, the 2018 Farm Bill provides for consideration of industrial hemp research proposals under federally-funded programs, like those within the United States Department of Agriculture. We note that effective in August 2018, a new, Hatch multi-state project funded and sponsored by the USDA National Institute for Food and Agriculture (USDA-NIFA) was officially established. This is definitely a positive step toward federally-supported research with industrial hemp.

A fourth consideration for producers and processors is: **Will federal policy and regulatory changes eventually allow the commercial production of hemp in the United States and what will be the role of the U.S. Drug Enforcement Agency (DEA), Food and Drug Administration (FDA) and other federal agencies in regulating future production and marketing of hemp and hemp products?** The uncertainty surrounding future laws and regulations creates significant risk for those investing in the hemp industry at all stages in the marketing chain while limiting investment by other entrepreneurs. Legalizing commercial hemp production across the United States, without any supply control measures in place, will likely lead to periods of excess production and price volatility similar to what is experienced with other agricultural commodities. Given the controversial nature of the crop and budget pressures facing traditional U.S. crops, it appears unlikely that U.S. farm policy will include programs that would benefit U.S. hemp producers. Also, changes in U.S. policy and regulatory controls will not occur in a vacuum as other nations adjust their policies, which will also affect the competitiveness of U.S. hemp production and hemp products.

The fifth consideration for producers is: **Where will the hemp industry in the United States develop?** With 40 U.S. states currently positioning themselves for this niche but growing market, it is not practical to expect (unless product demand grows substantially) that the market can sustain viable hemp production in every state. Hemp processors will not likely locate in every state and considering transportation costs (especially for fiber) and access to markets, technology may dictate that production will ultimately be concentrated in relatively few states where hemp can be grown at the lowest cost of production and transported shorter distances for processing. This suggests that states that can entice processors, manufacturers, and infrastructure to locate in their state based on a strong research base of knowledge, an interested, willing, and educated grower pool with lowest cost of production for desired quality characteristics requested, along with support from local and state governments will likely enhance their chances for success in this emerging industry.

A final consideration for producers and processors is: **Will consumer demand for hemp products continue to expand at an escalating pace?** A large portion of emerging hemp product sales in the United States has arguably been made by a few niche retailers marketing to a niche consumer base. As the market matures, will the "economics of hemp" be attractive enough for large retailers and input supplies to make hemp products available to and purchased by a larger consumer base to rapidly expand production opportunities for hemp farmers? This is an extremely complicated question, and currently, at the consumer level, we have not collected the needed information to provide a complete picture of consumer demand for products that contain hemp. What we have been able to determine is that the hemp products available across the United States vary widely. For example, one state may have access to hemp nuts, where another does not. For the industry to continue to grow, product availability will have to continue to expand.

Summary

All things considered, it is an exciting time in the United States for the production of industrial hemp. The crop that has been produced for thousands of years. Ithe United States has been illegal to produce for almost six decades, but became semi-legal to produce again in 2014. The passage of the 2014 Farm Bill provided the ability for producers, under strict regulatory control, to bring this crop from 0 acres of production to over 25,000 acres in under five years. Additional expansion will likely occur in the early years of the 2018 Farm Bill. To date, forty states have passed legislation to provide for and regulate the production of hemp. A critical question for the foreseeable future will be can anticipated demand for hemp-derived products continue to outpace expected increases in hemp production in the United States and possibly elsewhere? Significant price volatility could evolve, depending on the supply and demand balance. Also will the industry grow at the pace to encourage additional infrastructure in multiple states that might have been stifled with all the political uncertainty. As with any new product or crop entering the market, there have also been setbacks and hurdles, and with hemp, there are still many uncertainties that have yet to be addressed.

Global hemp production had declined significantly from its peak in the 1950s, but has been rebounding over the past decade in response consumer demand and policy changes. Just in the United States alone, the sales of hemp products has grown by double digits from 2012 to 2016. One of the rapidly growing areas that have been a significant contributor to this growth has been the production of CBD. Cannabidiol production has been the focus of Kentucky hemp producers and many of the 1400 plus hemp license holders in the United States. The primary reason for this is that expected returns for CBD are significantly higher than either grain or fiber production. However, the costs for its production are significantly higher, and the uncertainties regarding whether the federal government will begin to increase regulation is still unknown. Under the current regulatory and production climates, these issues make CBD production extremely risky for U.S. farmers, but also potentially very profitable, especially for early entrants in this market who contract with reputable processors with a sound business plan.

Hemp grain and fiber markets are expected to compete with the traditional crops being produced across the United States. Therefore, to grow in acreage, hemp crops will have to produce expected returns that are comparable to the crops a producer is currently growing. As shown in Table 1, hemp grain expected return is only slightly positive, and hemp fiber has a negative expected return above variable costs. However, hemp is increasing in competitiveness as improved genetics are available and domestic seed production is established, both of which will increase the yield potential. Finding ways to increase the competitiveness of United States hemp producers is going to be imperative given that other countries are also considering the production of industrial hemp. Many producers and researchers are also still experimenting with crop rotations and ways to manage weed pressure because no herbicides are labeled for use in hemp. The passage of the 2018 Farm Bill could significantly change this situation if hemp is removed from the controlled substance list and chemical companies begin to add hemp to the label. This would provide a significant boost to yields and eliminate one of the largest contributors to crop failure that hemp producers are facing currently; uncontrollable invasions of weeds.

Passage of the 2018 Farm Bill potentially eliminates a number of other hurdles that producers and processors are currently dealing with. Specifically, this legislation could potentially open the door allowing lending institutions to loan money for the production of hemp. It could also allow for the development of risk management tools such as crop insurance for producers to utilize. It will take time to develop these products, and given the federal budget constraints, it will not be simple. Lastly, federally-funded hemp research programs will likely evolve with the passage of the 2018 Farm Bill. This should increase research efforts and our body of knowledge significantly and rapidly.

Even though it is an exciting time in the United States for hemp producers, logic calls for cautious optimism and concurrent, strong efforts to maintain the sound financial position of their agricultural enterprises. Hemp is just another crop that can be produced in a wide range of locations across the globe, and is not a silver bullet solution to any part of our agricultural economy. The one potential exception to this is CBD, but today that can be a significant gamble. In general, hemp should only be produced if it has a positive impact on the farm's bottom line. Additionally, given the current environment and uncertainties, a producer needs to be able to lose 100% of their investment in growing this crop and be able to survive without long-term consequences.

References

AgCanada. 2018. Hemp acres, prices down for 2018. AgCanada, Glacier Media Group, Vancouver, B.C. https://www.agcanada.com/daily/hemp-acres-prices-down-for-2018 (Accessed 13 Mar. 2019).

Amason, Robert. 2017. Manitoba hemp processor doubles contracts. The Western Producer. Glacier Media Group, Vancouver, B.C.

Carus, M., and L. Sarmento. 2016. The European hemp industry: Cultivation, processing and applications for fibres, shivs and seeds. Vol. 2003. European Industrial Hemp Association, Hürth, Germany. http://eiha.org/media/2016/05/16-05-17-European-Hemp-Industry-2013.pdf (Accessed 13 Mar. 2019).

European Industrial Hemp Association. 2017. Record cultivation of industrial hemp in Europe in 2016. Press Release. European Industrial Hemp Association, Hürth, Germany.

FAOSTAT. 2017. FAOSTAT Data. Food and Agriculture Organization of the United Nations, Rome, Italy. http://www.fao.org/faostat/en/#home (Accessed 13 Mar. 2019).

Halich, G. 2018a. Corn, soybean and wheat budgets. University of Kentucky Cooperative Extension, Lexington, KY.

Halich, G. 2018b. Custom machinery rates applicable to Kentucky. University of Kentucky Cooperative Extension, Lexington, KY.

Hemp Business Journal. 2017a. European hemp cultivation grows by 32% to 33,00 ha in 2016. Market Report. Hemp Business Journal, Denver, CO.

Hemp Business Journal. 2017b. U.S. hemp cultivation acreage. Hemp Business Journal, Denver, CO. https://www.hempbizjournal.com/u-s-hemp-cultivation-acreage-data-charts/ (Accessed 13 Mar. 2019).

Hemp Business Journal. 2018. The U.S. hemp industry grows to $820 million in sales in 2017. Hemp Business Journal, Denver, CO. https://www.hempbizjournal.com/size-of-us-hemp-industry-2017/.

Hemp Industry Daily. 2018. Canada hemp changes could disrupt global CBD markets, but U.S. producers hopeful. Hemp Industry Daily, 16 October. https://hempindustrydaily.com/canada-hemp-changes-could-disrupt-global-cbd-markets-but-us-producers-hopeful/ (Accessed 13 Mar. 2019).

Johnson, R. 2018. Hemp as an Agricultural Commodity. Congressional Research Service Report. Congressional Research Service, Washington, D.C. https://www.votehemp.com/wp-content/uploads/2018/08/CRS-RL32725-Hemp-2018-06-22.pdf (Accessed 13 Mar. 2019).

Nichols, Kristen. 2017. Flower power: Canada's new hemp rules boost CBD production, but limits remain. Hemp Industry Daily, 29 November.

Syngenta Canada. 2017. Canadian hemp area expected to rebound in 2017. Syngenta Canada, Guelph, ON. https://www.syngenta.ca/News/market-news/canadian-hemp-area-expected-to-rebound-in-2017 (Accessed 13 Mar. 2019).

Vote Hemp. 2017. 2017 state hemp legislation. Vote Hemp, Washington, D.C. https://www.votehemp.com/wp-content/uploads/2018/09/Vote-Hemp-2017-US-Hemp-Crop-Report.pdf (Accessed 13 Mar. 2019).

Epilogue

D.W. Williams, PhD

These are truly exciting times for those in the U.S. who have supported utilizing hemp and hemp-derived products for many years. Recent and real efforts to legalize hemp production and utilization in the U.S. are easy to trace back to the 1990s. While it has taken 20 or more years to accomplish, passage of the 2018 Farm Bill has removed hemp from the Controlled Substance Act, and at the very least, has opened the door wide for production and utilization of hemp grain and fibers. It's exciting for us newcomers, too.

The resulting level of public interest in hemp is phenomenal. Those of us working in agriculture at U.S. land-grant universities have never experienced anything like this. Interest in the crop in the form of questions submitted by email, telephone calls, and visits to our offices or research stations has been literally unprecedented. The numbers of people participating in hemp research field day events are always in the hundreds, and include a very diverse audience ranging from literally no experience at all in agriculture but exhibiting sincere and strong interest in forwarding efforts to grow hemp, all the way up to the most successful and productive farmers in the state; those farming tens of thousands of acres, and every level of experience in between. We don't see very many of these same people at corn or soybean field day events. It is totally clear that this is a hemp-based phenomenon.

Crop scientists like me tend to be rather pragmatic (but certainly not always; like any population of humans, there's also true diversity among agronomy geeks). This is especially true considering old geeks like me. We tend to rely very heavily on logical thinking and usually on data or known facts to support conclusions in all facets of our lives; professional and personal. Our learning curve as university scientists tasked with working with hemp as a brand new crop was very steep, indeed. We immediately began relying on both the scientific literature and personal communications with hemp researchers across the globe in order to inform ourselves appropriately and manage productive and efficient domestic research efforts. Well, it didn't take long at all to realize that most of this was really not new. Of course, humans have utilized oilseeds and natural fibers for their entire existence, and hemp was really just one of several potential sources of both. In short and excepting the genus name to which the species belonged, it became apparent very quickly that in many respects, there's really not much that's spectacular about hemp grain and fiber. We can today, always have been, and continue to utilize other species in the same spaces that hemp can contribute to today. This is not a negative, anti-hemp statement. Only a pragmatic one

While I am not at all trained in understanding the human mind or social science at any level, one thing is still very clear. The public interest in hemp today is extremely broad and diverse, even if it's not entirely supported by science. Consider that my colleagues at the University of Kentucky (UK) and others all across the country have been working in the oilseed and natural fiber spaces for essentially always. Where was everyone a few years ago when two of my colleagues released a new variety of chia (Salvia hispanica L.); one that was very well-adapted to our latitude and climate with high grain yields? Chia seeds are also regarded as very healthy in many of the same respects as hemp seeds. We're still growing chia selection plots on our experiment station today. Are any of you aware of that release? I'm guessing not. As far as I know, only one farmer in Kentucky is producing the seed.

Strongly related; where was everyone in the late 1980s and early 1990s when another colleague at UK began conducting both variety and general agronomy trials with kenaf (Hibiscus cannabinus L.). This work even elicited a visit to the experiment station by the state police once, as the morphology of hemp and kenaf can be quite similar. This was great work towards identifying high-yielding varieties for Kentucky and defining basic agronomic parameters to optimize yields. I have referred to that work several times in designing experiments comparing hemp to kenaf today. Basically, except for my colleague (since deceased) and a few others in our department, there was essentially zero interest in that work, so it faded away rather quickly and completely. It is well-known today that bast and hurd fibers from kenaf are totally adequate for many of the same applications as are hemp fibers. I don't think they ever had a kenaf field day, but honestly I don't remember. All of this interest in hemp alongside total ignorance of the other species that can serve in the exact same roles indicates one thing: for whatever reason, hemp is much, much cooler, and hence almost certainly much, much more marketable, even when its performance is not superior. This fact alone supports moving forward posthaste with research work to support the evolving industry. Which car would you buy; this car contains kenaf parts, this car contains hemp parts? It is nearly a no-brainer. Most will ask "what is kenaf"? This broad and real connection to hemp by the general public both in Kentucky and well beyond challenges my natural pragmatism, but it also fuels my motivation to work hard towards contributing to the positive evolution of the hemp industry today. Bottom line: hemp is cool. People will buy it.

I still chuckle today when people refer to the wonderful efforts of Henry Clay, the "Great Compromiser", towards the very successful hemp industry in Lexington, Kentucky in the early- to mid-19th century. The original paperwork from his farm, Ashland, still exists today and refers often to his hemp crops. It is clearly a matter of modern and legal semantics, but the probability that Senator Clay grew cannabis with 0.3% or less THC measured in the floral material is probably near zero. In other words, he and others of that era almost certainly grew dope for rope, yet these efforts are often touted when the history of hemp is considered or forwarded in support of modern efforts. Our collective connection to cannabis today is absolutely a real thing, and often exists regardless of the levels of THC.

The cannabinoids are different. While not totally exclusive to the species, we are not aware of any other species that produce cannabinoids at the same levels as Cannabis sativa L. There's so much we don't know about the utilization of these molecules, but the same or even higher levels of public interest exist in this space, too. I now have direct, collaborative relationships with colleagues in both the colleges of pharmacy and medicine at UK. This is both new to me and extremely exciting. How many old agronomists get to work with a brand new (to our generations) plant species as a potential crop to produce molecules that may contribute a great deal to multiple facets of modern medicine? How about none? Today, the cannabinoid industry anxiously awaits guidance from the U.S. federal government regarding a regulatory framework for the cannabinoid molecules. Regardless of the ultimate regulatory framework, it seems clear that research will continue and expand investigating potential utilizations for cannabinoids. Since they are derived from plants, we will concurrently work to optimize efficient production protocols. These are exciting times, indeed.

Our intentions in providing this book were two-fold. First, we hope the work will be useful to anyone interested in producing, processing, or utilizing natural fibers, grain, and ultimately cannabinoids from hemp today. We've endeavored to provide current information addressing all of those interests. Secondly, we hope it could be useful later in time, perhaps as a snapshot taken early in the evolution of the modern hemp industry in the U.S. Maybe future authors might refer to this work as they make "way-back-when" statements in future writings. As such, we've also tried to provide current and accurate information on the industry in general as it exists today. These are indeed very exciting times. Rock on.

Printed and bound by CPI Group (UK) Ltd, Croydon, CR0 4YY

27/10/2024

14580340-0002